AF841680

CONTENTS

NEW TRENDS IN ECOLOGY RESEARCH

NEW TRENDS IN ECOLOGY RESEARCH

A. R. BURK
EDITOR

Nova Science Publishers, Inc.
New York

For permission to use material from this book please contact us:
Telephone 631-231-7269; Fax 631-231-8175
Web Site: http://www.novapublishers.com

LIBRARY OF CONGRESS CATALOGING-IN-PUBLICATION DATA:

New trends in ecology research / A. R. Burk, editor.
p. cm.
Includes index.
ISBN 1-59454-379-8 (hardcover)
1. Ecology--Research. I. Burk, A. R.

QH541.2.N49 2005
577'.072--dc22 2005005416

Published by Nova Science Publishers, Inc. ✣ New York

PREFACE

Ecology is the study of the interrelationships between organisms and their environment, including the biotic and abiotic components. There are at least six kinds of ecology: ecosystem, physiological, behavioral, population, and community. Specific topics include: Acid Deposition, Acid Rain Revisited, Biodiversity, Biocomplexity, Carbon Sequestration in Soils, Coral Reefs, Ecosystem Services, Environmental Justice, Fire Ecology, Floods, Global Climate Change, Hypoxia, and Invasion. This new book presents new research on ecology from around the world.

Chapter 1 discusses the key role ozone plays in global atmospheric chemistry and is an important greenhouse gas. The surface ozone is a secondary pollutant formed from the precursors NO_x and VOCs during the photochemical reactions. The meteorological situation is very important for the concentration levels. The hot sunny stabile weather is conducive for surface ozone formation. The surface ozone is a pollutant of major concern due to its impacts on receptors at currently occurring ambient levels in many regions of the world. The ambient air pollution, resulting mostly from energy production, belonged to the major environmental problems in former Czechoslovakia between 1960s and 1980s. As a result of profound political and economic changes in Central Europe in 1989, the air pollution (esp. the concentrations of SO_2 and SPM) was significantly reduced. The photochemical smog, the indicator of which is the surface ozone, nevertheless, remains a hot issue.

Chapter 2 is divided into three main sections. The first section is dealing with uncertainties in empirical data, defining and discussing e.g., the highest r^2 of predictive models, the sampling formula and uncertainty in empirical data, empirically based highest r^2, highest reference r^2, CV and sampling period, CV versus mean values, mean versus maximum values and spurious correlations. The second section includes issues concerning model testing, i.e., calibration and validation, sensitivity tests and uncertainty tests using Monte Carlo techniques. The third section introduces some basic ideas concerning dynamic mass-balance models for lakes. Algorithms for water mixing and outflow are given and efficient ways to define deep water, surface water, erosion and accumulation areas are discussed. This section also discusses time and area resolution in dynamic lake models and the most efficient way of utilizing distributed empirical data.

Since the development and popularisation of statistical science during the 1930s, the uptake of statistical hypothesis tests in major science disciplines like psychology, medicine and biology has been rapid. Hypothesis tests did not become common in the ecological literature until the 1960s but, since then, testing zero null hypotheses and using P-values to make dichotomous decisions about rejecting them has become routine. Recent surveys of the ecological literature show that P-values are reported in the vast majority of published research articles. The authors of chapter 3 believe that the focus on tests of zero null hypotheses and the reliance on P-values is inappropriate and argue that ecological data are more relevant to both ecologists and applied users if *a priori* sample size determination procedures are conducted, and effect size estimates and their associated confidence intervals are used to present results. Chapter 3 reports on evidence that effect sizes, associated uncertainty estimates and sample size calculations are missing from a large number of publications in two leading conservation biology journals. An overwhelming majority (92%) of surveyed articles reported P-values, but only 3% of these reported statistical power. Furthermore, many articles were missing at least one estimate of effect size (43%) or measure of variance (67%). Confidence intervals were reported in only 19% of articles and, when reported, were often inadequately interpreted.

According to chapter 4, the correct understanding of ecological and environmental processes is indispensable for the sustainable development of Humankind. For practical purposes, the account of ecosystem complexity is greatly aided by the application of computer aided technologies. This paper presents a methodological approach linking computer modelling of natural processes with geoinformation technology, realised in the PC modelling System "Relief-Processor" (R_P). The paper also delivers a word of warning as regards possible pitfalls with application of certain computer technologies for modelling of ecological processes, giving an example of errors which may occur while using Microsoft Excel for fitting trendline equations.

Taking a glance at chapter 5 one may see that an increase in the consumer's willingness to pay (WTP) for an improvement in the ecological performance of a marketed product should be welfare improving. Furthermore, it should create the appropriate incentives for firms to get involved in such improvements by benefiting "green" producers and harming ecologically indifferent ones. We provide some theoretical arguments against these apparently obvious affirmations, showing that the profitability and welfare effects of WTP increases for green products depend on whether they are targeted towards the least ecologically aware of the consumers. The authors hypothesis concerning these different kinds of changes in ecological consciousness is supported by Spanish survey data on subjective perceptions of the consequences of ecological consciousness-increasing advertisements. The results show that, according to the authors respondents, the socially undesirable type of messages represents a 40% of the total, calling for a more cautious attitude towards qualitative aspects of state advertising aimed at creating a more ecologically aware consumer.

Matrix models, i.e. expressing systems of linear equations in matrices and using matrix algebra, have been successful in describing the demographic dynamics of a population. The same method can be used to study the effect of space on population growth and spatial distribution. Here, a temporally and spatially discrete matrix model is used to ask general and specific question in landscape ecology. One such question is how habitat amount and arrangement affects population distribution and persistence, for example how habitat loss in various habitat arrangements (random, clumped or fragmented landscapes) affects a population. Chapter 6 reviews how matrix models have been used to determine the effect of spatial structure of a population and how this approach have been generalised to a multi-patch system. The authors demonstrate that a matrix model (including population redistribution, survival and growth) can serve as a general framework for spatial redistribution in a landscape. The results are comparable with existing results from studies of critical thresholds. Existing analytical tools are implemented to determine the sensitivity of e.g. population growth and stable population distribution (eigen-value and -vector sensitivity) to changes in transitions between landscape cells. Some examples and illustrations of how sensitivity values can be used in landscape management are provided.

The use of indices has expanded into several natural resource disciplines, including ecological studies, environmental policymaking, and agricultural economics. Nevertheless, despite their increasing use in natural resource disciplines, researchers and public decision makers continue to express concern about validity of these instruments to capture and communicate multidimensional, and sometimes disparate, characteristics of research data and stakeholder interests. The authors purpose in chapter 7 is to demonstrate how useful indices can be for communicating environmental information to decision makers. We discuss how environmental indices have evolved from simple to multicriteria formats, and then propose two advancements, the impact matrix and disparate stakeholder management.

The decisions of many individuals and social groups, taking according to well-defined objectives, are causing serious social and environmental problems, in spite of following the dictates of economic rationality. There are many examples of serious problems for which there are not yet appropriate solutions, such as management of scarce natural resources including aquifer water or the distribution of space among incompatible uses. In order to solve these problems, chapter 8 first characterizes the resources and goods involved from an economic perspective. Then, for each case, the chapter notes that there is a serious divergence between individual and collective interests and, where possible, it designs the procedure for solving the conflict of interests. With this procedure, the real opportunities for the application of economic theory are shown, and especially the theory on collective goods and externalities. The limitations of conventional economic analysis are shown and the opportunity to correct the shortfalls is examined. Many environmental problems, such as climate change, have an impact on different generations that do not participate in present decisions. The chapter shows that for these cases, the solutions suggested by economic theory are not valid. Furthermore, conventional methods of economic valuation (which usually help decision-makers) are unable to account for the existence of different generations and tend to obviate long-term impacts. The chapter analyzes how economic valuation methods could account for the costs and benefits enjoyed by present and future generations. The chapter studies an appropriate consideration of preferences for future consumption and the incorporation of sustainability as a requirement in social decisions, which implies not only more efficiency but also a fairer distribution between generations than the one implied by conventional economic analysis.

In: New Trends in Ecology Research
Editor: A. R. Burk, pp.1-36
ISBN: 1-59454-379-8

Chapter 1

SURFACE OZONE: TEMPORAL AND SPATIAL PATTERNS AND POTENTIAL IMPACT ON ECOSYSTEMS IN THE CZECH REPUBLIC AS INDICATED BY EXPOSURE INDEX AOT40

Iva Hůnová, Hana Livorová and Jana Ostatnická
Czech Hydrometeorological Institute Prague, Czech Republic

ABSTRACT

Ozone plays a key role in global atmospheric chemistry and is an important greenhouse gas. The surface ozone is a secondary pollutant formed from the precursors NO_x and VOCs during the photochemical reactions. The meteorological situation is very important for the concentration levels. The hot sunny stabile weather is conducive for surface ozone formation. The surface ozone is a pollutant of major concern due to its impacts on receptors at currently occurring ambient levels in many regions of the world.

The ambient air pollution, resulting mostly from energy production, belonged to the major environmental problems in former Czechoslovakia between 1960s and 1980s. As a result of profound political and economic changes in Central Europe in 1989, the air pollution (esp. the concentrations of SO_2 and SPM) was significantly reduced. The photochemical smog, the indicator of which is the surface ozone, nevertheless, remains a hot issue.

The surface ozone concentrations over the Czech Republic have been measured since 1993. The network has grown denser gradually and 60 stations are under operation currently. These represent rural, mountainous and urban areas and record ozone concentrations continuously giving 30-min means. The ozone concentration measurement is based on ultraviolet absorption photometry.

The chapter presents results of 10-year surface ozone monitoring over the Czech Republic. Daily and seasonal variability of concentrations and trends for the different types of stations are presented. The data set is analyzed as to their temporal and spatial patterns. Moreover the exceedances of limit values - 24-h mean concentration of 65 $\mu g.m^{-3}$ (a threshold value for vegetation protection) and exceedances of 8-h mean concentration of 120 $\mu g.m^{-3}$ (a limit value for human health protection) – are assessed.

The AOT40 exposure index developed within the context of the UN/ECE as a critical level, a tool commonly used in Europe to assess the geographical areas with vegetation potentially at risk due to elevated ambient ozone concentrations, is calculated. The AOT40 concept is discussed and compared to the more sophisticated flux approach. The maps of AOT40 indicate, that critical levels both for forests and crops/semi-natural ecosystems are exceeded over great portions of the Czech territory throughout the period under review, even in years with meteorological conditions unfavorable for ambient ozone formation.

The experience with passive samplers used for ozone measurements in mountainous-forested areas and suitability of the method for ecology research is also discussed.

INTRODUCTION

Ozone in the Atmosphere

Ozone (O_3) plays a vital role in global atmospheric chemistry (Colbeck and Mackenzie, 1994; Seinfeld, 1996; Warneck, 1988). It is one of the most important indicators of the oxidising power of the atmosphere. Due to their key role, oxidation reactions are sometimes referred to as nature's "cleansing" process, and the overall rate of this process as the "oxidation capacity" of the atmosphere (Prinn, 2003). Photolysis of O_3 in the presence of water vapour produces the OH radical, which is the main atmospheric oxidant responsible for removal of a large number of important gases (Brasseur, 2003). Without OH the Earth's atmosphere would have a different composition, as the present trace gases would dominate it.

O_3 is a powerful infrared absorber and emitter, a greenhouse gas, and many other greenhouse gases, notably CH_4, are destroyed principally by reactions with OH. This means that there are significant connections between atmospheric processes and climate. Added to these greenhouse related connections is the fact, that the first step in the production of aerosols from gaseous SO_2, NO_2 and hydrocarbons is almost always reaction with OH. Aerosols play a key role in climate through absorption and/or reflection of solar radiation (Prinn, 2003).

Traditionally, when considering the role of atmospheric ozone, stratospheric and tropospheric ozone have been considered separately, even though the importance for both regions of what happens in the upper troposphere and lower stratosphere has well been recognized. Although ozone distribution and changes, and ozone-climate links have partly been included and discussed in the most recent assessment (IPCC, 2001; WMO, 2003), no thorough scientific overview of this problem is currently available (EC, 2003).

About 90 % of O_3 resides in the stratosphere, where it performs a vital role as a shield against low wavelength ultraviolet light harmful to many life forms. It is formed from the direct photolysis of oxygen with a maximum wavelength of light of 242 nm. Such strong UV radiation does not enter the troposphere.

Mere 10 % of Earth's ozone is present in the troposphere (the lowest part of the atmosphere, extending from the surface to 10–15 km altitude). Earlier theories concerning the origin of ozone in the troposphere proposed a purely physical picture with the stratosphere as the source and gaseous deposition to the surface as the sink. A more complex one in which chemistry also plays a major role has now superseded this. The main sources of O_3 in the troposphere are *in situ* photochemical production from photolysis of NO_2 assisted by the HO_2

– OH chain reactions, and photochemical destruction from ozone photolysis and the same chain reaction. The switch from net destruction to net production is determined mostly by the levels of NO_x - NO and NO_2 (Penkett et al., 2003). The distribution of O_3 in the troposphere is controlled by few main processes: emission of ozone precursors, *in situ* photochemical processing, and transport within the troposphere. The tropospheric ozone has a certain vertical distribution of concentrations.

Surface ozone, a part of tropospheric ozone column most adjacent to the Earth's surface, is a pollutant considered at present as one of the most hazardous species due to its toxic effects and ambient levels exceeding the limit values on vast territories in Europe, North America and Asia. Adverse impacts of ozone on human health, vegetation and ecosystems and materials are widely recognised.

Surface Ozone

Surface ozone is an indicator of the photochemical air pollution often referred to as the photochemical smog. The photochemical smog, first identified in Los Angeles in 1940s, is now a widespread phenomenon in many of the world's big cities. First this problem was associated exclusively with cities, since 1970s it is known as a widespread regional problem. Considerable progress has been made in identifying chemical processes responsible for generation of O_3 and other secondary photochemical pollutants in the planetary boundary level. This has been achieved by a combination of field observations, laboratory investigations and numerical modelling studies.

Ozone is a secondary pollutant formed from the precursor compounds (NO_x and VOCs) in complicated photochemical reactions. Due to these reactions it is not a straightforward task to decrease the ambient concentrations, as it was e.g. in case of SO_2 or TSP. The complex chemistry leads to the formation not only of O_3, but a variety of additional oxidising substances, such as e.g. well-known peroxyacetylnitrate, PAN ($CH_3C(O)OONO_2$). Normally hydrocarbons are the limiting factor in the photochemical O_3 production in the city centres, and NO_x is the limiting factor away from urban centres (Sillman, 1999), in most of the non-urban troposphere (Crutzen, 1988).

Ozone is produced directly by photolysis of NO_2:

$$NO_2 + h\nu \rightarrow NO + O \qquad (\lambda < 400 \text{ nm}) \tag{1}$$

where the oxygen atom (O) rapidly recombines with molecular oxygen (O_2) to produce ozone (O_3). The photolysis of ozone is the sole known anthropogenic source of tropospheric ozone (Singh and Fabian, 2003). Normally this reaction is counterbalanced by the reaction of NO with O_3:

$$NO + O_3 \rightarrow NO_2 + O_2 \tag{2}$$

Each of these reactions occurs very rapidly, on a time scale of 200 s or less (Sillman, 2002).

Typically NO and NO_2 adjust to establish a near-steady state between reactions (1) and (2). There are two distinct situations, however, when the above reactions result in a net change in O_3 concentration: removal of O_3 via reaction (2) at night time or in the vicinity of large NO_x sources and ozone production associated with daytime NO_x–VOC–CO chemistry.

Ozone concentrations undergo distinct seasonal and diurnal variations. Peak concentrations are recorded on afternoons with intense sunshine in spring and summer.

The highest O_3 levels in rural areas are recorded downwind of urban plumes. The small-scale variations are dependent on the vicinity of NO emission sources. In urban areas, reduced O_3 levels are observed in the vicinity of traffic arteries, in rural areas, O_3 levels are more or less uniformly distributed.

According to Percy et al. (2003) three types of trends in surface ozone concentrations are apparent: (1) a huge increase in the extent of O_3 and the forest areas at risk; (2) a decrease in maximum 1-h O_3 concentrations, at least in the northern hemisphere countries with existing O_3 precursor control programs in place; and (3) an increase in background O_3 concentrations over much of the world.

Surface Ozone and its Effects

Ozone is a strong oxidant with significant negative impacts on human health, vegetation and ecosystems (Ashmore, 2003), materials and constructions. The effects are documented for current ambient levels. The increasing surface ozone ambient concentrations as well as extending spatial distribution have already raised concern in Europe (Stanners and Bourdeau eds., 1995; EEA, 1998; EEA, 2003) and the USA (US EPA, 1996; Chappelka et al., 1999) for a longer time. Recently the surface ozone has been considered as a global issue (Ashmore, 2003), as the number of evidences on its negative effects in Asia, Africa and Southern America grows (Emberson et al., 2001).

The scientific consensus recognizes ambient air pollution as a predisposing or triggering factor to forest decline (Muzika et al., 2004). The deposition of ozone is hypothesised to contribute to the impairment of forest health status in the vast areas in Europe (Coe et al., 1995) and North America (McLaughlin and Kohut, 1992). Worldwide, Fowler et al. (1999) estimate that some 24 % of the world's forests are currently exposed to damaging concentrations of O_3 and that this number will increase to 50 % of the world's forests by the year 2100.

From the field studies/experiments is concluded, however, that in most of the regions where the forest decline has been detected, there are several likely causes of the damages. The conifers in higher altitudes are exposed to the cloud water containing the high concentrations of major ions SO_4^{2-}, NO_3^-, NH_4^+ and H^+ in the long run (Mohnen, 1992). Moreover there is an influence of the occult/horizontal precipitation with major ion concentration manifold higher than routinely measured in rain or snow (Crossley, 1992). The higher altitude regions are also influenced by the elevated concentrations of surface ozone due to the occurrence of stabile nocturnal boundary layer suppressed by the higher wind speed (PORG, 1997). The result of experimental studies in laboratories and open-top chamber nevertheless clearly prove the O_3 impact on physiological processes and growth of the range of common tree species of the temperate zone (Skarby et al., 1998; Chappelka and Samuelson, 1998).

The ozone impact on forests is even more serious when it coincides with sulphur and nitrogen oxides (Pasuthová et al., 1987; Mészáros, 1993). The importance of pollutant mixtures in determining plant responses and their importance to calculate reliable critical levels has been clearly recognized. With the existing experimental methodology it is difficult, however, to define reliable critical levels for pollutant combinations (Harrison ed., 1999; Fangmeier et al., 2002).

The rapidly changing atmospheric environment with its mix of increasing anthropogenic emissions means that the future world's forests will be faced with unprecedented levels of both greenhouse gases and ambient air pollutants (Karnosky et al., 2003).

Surface Ozone – Impacts on Vegetation and Ecosystems

Due to its toxicity, spatial distribution and frequent occurrence the surface ozone is the most common oxidant phytotoxic substance in ambient air (Fishman, 1991). The effects of ozone on the biochemical, cellular and physiological levels are known (Long and Naidu, 2002). The main effects on terrestrial vegetation include the macroscopic damages (Manning, 1993), reduction in photosynthesis, growth and productivity (US EPA, 1996), change in diversity of plant communities (Barbo et al., 1998) and increased sensitivity toward both the abiotic and biotic stress factors (Chappelka and Chevone, 1992).

Like other gases within the atmosphere, O_3 is taken up by plants through stomatal diffusion (Hill, 1971). There is a general agreement that stomatal conductance, determined both genetically and environmentally, is often the major determining factor of plant sensitivity to ozone. Stomata tend to close partially when the atmosphere is dry or when the water content of the soil is low (Harrison ed., 1996). Differences in O_3 sensitivity are often well correlated with such changes (Alscher and Wellburn eds., 1994). The essential role of the influence of water supply on stomatal activity has been found in some experimental studies (e.g. Cieslik, 2004).

Apart from the stomatal uptake, in some cases the wax cuticule covering the leaves can act as a photochemical sink, esp. under high solar radiation (Rondón, 1993). The toxic effects of this path remain still unclear; the important consequences for coniferous forests under the summer anticyclone meteorological situations in Europe when ambient ozone levels can increase substantially are assumed, however (Coe et al., 1995). The non-stomatal flux can, according to some studies (Zhang et al., 2002) even dominate over the stomatal flux (Fowler et al., 2001).

Referring to effects, the daily concentrations of ozone are mostly taken into account. The night concentrations are mostly neglected due to the assumption that ozone concentrations and conductance are minimal during the night. The ozone levels can, however, stay relatively high even during the night, specifically in mountain regions. Although the conductance tends to be lower in night than during the daytime for most of the plants, the conductance during the night can result in considerable ozone uptake. Moreover, the plants can be more sensitive towards the exposure during the night due to the protective mechanisms being less effective than during the daytime (Musselman and Minnick, 2000).

Many experimental studies aimed at elucidation of ambient ozone impacts on vegetation have been carried out in both the laboratory and field. From results of dose-response studies with open-top chambers it is concluded that mean concentrations are not appropriate to characterize ozone exposure. This is due to the fact, that effects of ozone resulting from cumulative exposures and higher concentrations have greater effects than lower ones (WHO,

2000). The "translation" of concentration patterns during the exposure period in a biologically meaningful way led to construction of exposure indices.

Assessment of Ozone Impact: Exposure Indices and Flux Concept

In Europe the current critical levels for O_3 to protect vegetation and ecosystems (crops, natural and semi-natural vegetation and forests) are based on the accumulated exposure over a threshold of 40 ppb (exposure index AOT40). There is a widespread agreement among plant effect scientists that AOT40 is a substantial simplification, as the crucial fact is the dose absorbed by the plant and not the concentration measured in ambient air. It is much easier, however, to measure the ambient ozone concentration needed for calculation of the exposure index than to measure the stomatal flux. That is the reason why the exposure index AOT40 approach is still used though considered as much less meaningful and scientifically sounded than the stomatal flux approach. At present the main problems with the flux approach, according to Grünhage et al. (2004), are attributed to uncertainties in the parameterisation of stomatal conductance, non-stomatal deposition and the representativeness of species used in the flux-effect studies.

In 1988 the United Nations Economic Commission for Europe (UN/ECE) adopted the critical levels concept. The critical levels were defined as "the concentrations of pollutants above which direct adverse effects on receptors, such as plants, ecosystems or materials, may occur according to the current knowledge" (UN/ECE, 1988). At subsequent workshops - held in Egham, UK in 1992 (Ashmore and Wilson eds., 1994), Bern, Switzerland in 1994 (Fuhrer and Achermann eds., 1994), Kuopio, Finland in 1996 (Kärenlampi and Skärby eds., 1996) and Gerzensee, Switzerland in 1999 (Fuhrer and Achermann eds., 1999) - the exposure index AOT40, defined as cumulative exposure over a threshold concentration of 40 ppb over a set up time period, was adopted and refined. A comprehensive review aimed at providing a description and critical assessment of the scientific basis for the critical levels for ozone AOT40 and identification of its advantages as well as limitations and uncertainties are given by Fuhrer et al., (1997).

The AOT40 corresponds to ozone exposures associated with significant negative effects on the yields of crops, the biomass increment of forest trees and species composition of semi-natural plant communities (Fuhrer et al., 1997). For crops, data on grain yield of spring wheat exposed in open-top chambers to different ozone concentrations over the growing season were used to establish the critical level, since that database is the largest and most consistent and wheat is agronomically important across Europe and it is known to be a sensitive species. For forests the database is small. The annual biomass increment in beech saplings has been used to establish the critical level (WHO, 2000). The appropriateness of using seedling responses to predict mature tree responses to ozone is, however, questionable considering the long life span of trees (Kelly et al., 1995).

Apart from AOT40, other alternative exposure indices for ozone are used as well (Lefohn et al., 1988; Finnan et al., 1997; Lissková, 1998). These are either cumulative indices with various threshold levels generally referred to as SUM, or indices giving greater weight to higher concentrations as W126 (an index derived from sigmoidally weighting ozone concentrations). Throughout the United States, the indices that are routinely employed include SUM0, SUM06, SUM08 and W126 (Panek et al., 2002 b). The AOT40 concept is most widely used in Europe where it is believed to be the best available approach in identifying geographical areas that may be potentially impacted by ozone (Ashmore and Fuhrer, 2000). It has been considered a fair compromise between sophistication, existing experimental data and practical needs (Fuhrer et al., 1997).

A disadvantage of the AOT40 exposure index is its high sensitivity resulting in stringent requirements for the quality of measured data. Inevitably uncertainties in both observed and calculated ozone concentrations and a high sensitivity of the ozone index results in instability of the exposure estimates and creates problems in its use (Sofiev and Tuovinen, 2001). Relative change in the ozone mean concentration appears as being multiplied by a factor of 2–10 in the value of AOT40, i.e. a 5 % overestimation in the mean concentration of ozone results in a 10–50 % overestimation in AOT40 (Tuovinen, 2000).

To date, a simple approach (referred to as Level I), using single values for three different types of receptors (forests, crops, semi-natural vegetation) neglecting the influence of different factors on plant response to ozone, has been used. Suitable methods for a more sophisticated Level II approach, the long-term goal of which should be a flux-based approach, are being developed (Fuhrer and Achermann eds., 1999).

The exposure index AOT40 has also appeared in the European Communities legislation ("acquis communitaire") in the EC third daughter directive on ozone 2002/3/EC (EC, 2002), and has been used as a long-term objective value targeted for vegetation and ecosystems protection. This has been transposed to national legislation of EU member states as well as accession countries and thus has become an important, mandatory tool of policy and decision-making.

The Ambient Air Quality Monitoring in the Czech Republic

The Czech Hydrometeorological Institute (CHMI) in Prague is the principal monitoring organisation in the Czech Republic. It runs a nation-wide monitoring network for ambient air quality. Other organizations (Public Health Service, Organization for the Rationalization of Power Plants, Czech Geological Survey, Forestry and Game Management Research Institute, Water Management Research Institute, Research Institute of Plant Production) and local authorities also contribute to the measuring activities.

CHMI operates two types of ambient air quality stations - automated and manually operated. The automated stations, a "backbone" system of the monitoring, measure continuously and give 30 min. mean concentrations while manually operated stations give 24h mean concentrations. The principal air pollutants recorded are SO_2, NO_x, aerosol particles (both TSP and PM_{10}), O_3, CO and heavy metals in aerosol. Chemical composition of precipitation is also monitored. Precipitation for chemical analysis is sampled on a monthly or weekly basis by automated wet-only or bulk samplers. All of the ambient air quality data recorded over the territory of the Czech Republic is stored in one central database ISKO (Air Quality Information System) run by CHMI and regularly used for ambient air quality assessment (Fiala and Ostatnická eds., 1995-2003). All institutes monitoring the ambient air over the Czech territory are encouraged to send the data to the central database. Moreover, the data measured in border regions of Germany and Poland is kindly sent into the database. This German and Polish data is very helpful for ambient air quality assessment in Czech border areas and is used for interpolation between measuring sites when constructing maps with concentration or atmospheric deposition fields. The individual measuring institutes are fully responsible for their own data quality control before sending the data into the database. The data gathered in the database is consequently verified using the standard several steps procedure centrally. The entire data set in ISKO is annually checked thoroughly comparing both the concentrations among neighboring stations and consecutive concentrations recorded at one site.

As a consequence of the political changes in former Czechoslovakia in the early 1990s, the EU directives have been adopted as a basis for environmental legislation, in particular relating to air quality. The Clean Air Act 86/2002 Co. forms the legislative framework for 10 implementing regulations. The range of pollutants monitored expanded, the new concept of limit value have been adopted. The ambient air quality limit values in the Czech Republic have aimed to protect human health so far. In line with the European Community (EC), new limit values for vegetation and ecosystems protection have been established in the Czech national legislation. The Czech ambient air quality network is under reconstruction at present in order to optimise the air quality measurements to fulfil the requirements of the new EC and Czech legislation.

The Aim of this Chapter

The aim of this chapter is to present the trends of surface ozone concentrations throughout the Czech Republic based on 10–year monitoring results and indicate the areas with ozone levels high enough to harm the vegetation.

METHODS

The Surface Ozone-Monitoring Network in the Czech Republic

The surface ozone concentrations over the Czech Republic have been measured since 1993. The network (Fig. 1) has grown denser gradually and 60 stations (Fig. 2) are under operation currently. These represent rural, mountainous and urban areas and record ozone concentrations as 30-min means. The ozone concentration measurement is based on ultraviolet absorption photometry. Some of the stations, within the meteorological monitoring framework, also record global solar radiation, important e.g. for calculation of ozone exposure index (see part 2.3).

For the data assessment the records from most of the ozone-monitoring stations operated by the Czech Hydrometeorological Institute are used. Table 1 presents the brief characteristics for sites selected for detailed presentation (daily or annual courses) in this chapter. The location of these is depicted at Fig. 1.

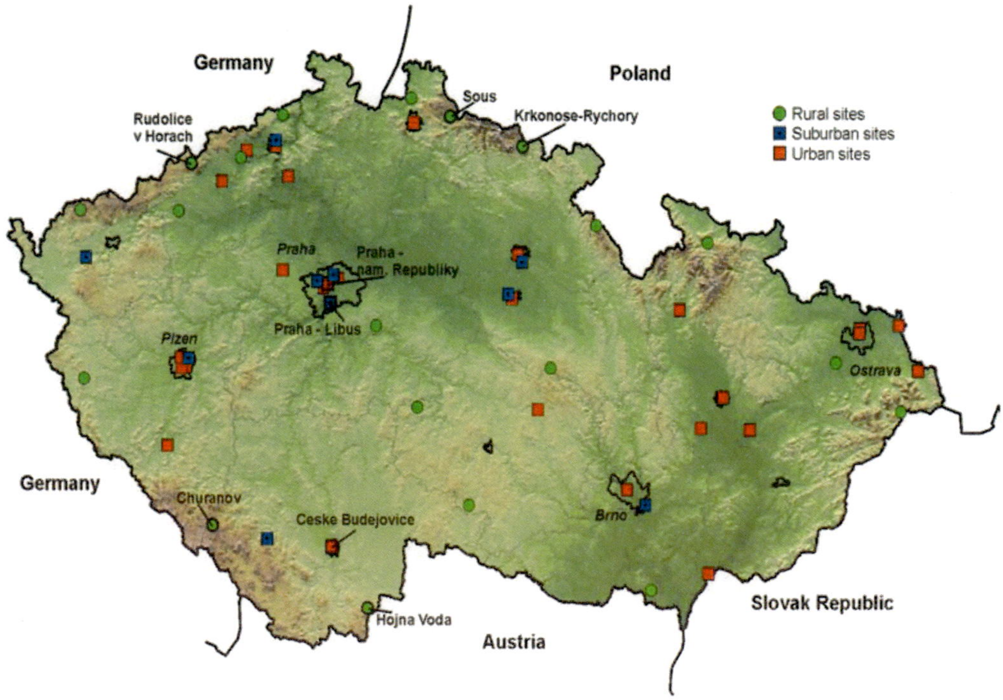

Fig. 1: The surface ozone-monitoring network, the Czech Republic, 2003 (the sites used for detailed analysis are indicated)

Fig. 2:The automated station monitoring ambient air quality in the Czech Republic, exterior and interior

Data Analysis

The input data for the trend analysis represent 30-min concentrations from 50 stations measuring at least 6 years. The histograms of concentrations for consecutive years have been plotted. The annual mean 50th and 98th percentiles were calculated and the data set was analyzed for the exceedances of 24-h mean concentration of 65 $\mu g.m^{-3}$, a threshold for vegetation protection (EC, 1992) and exceedances of 8-h mean concentration of 120 $\mu g.m^{-3}$, a limit value for human health protection (EC, 2002). The overall average for all stations was calculated, which minimized the local influences at different sites in consecutive years. To assure the comparability of single years, the concentrations from stations measuring during the entire period under review, were taken into account. The data coverage required was at least two thirds of the year's days. The analysis was carried out both for the entire data set and for the stations grouped as to the type of monitoring site (rural, urban, suburban, traffic and industrial); both for the entire year and for the summer and winter periods separately.

Table. 1: Selected surface ozone monitoring sites characteristics

Site	Area	Type of the station	Altitude m a.s.l.	Coordinates
Krkonose-Rychory	Krkonose Mts.	Rural/mountainous	1001	50°39′40.00" N 15°51′5.00" E
Churanov	Sumava Mts.	Rural/mountainous	1122	49°4′6.00" N 13°37′2.00" E
Hojna Voda	Novohradske hory Mts.	Rural/mountainous	818	48°43′26.00" N 14°43′33.00" E
Sous	Jizerske hory Mts.	Rural/mountainous	771	50°47′22.00" N 15°19′19.00" E
Rudolice	Krusné hory Mts.	Rural/ mountainous	840	50°34'52.41" N 13°25'16.98" E
Ceské Budejovice	Residential area in the town	Urban background	383	48°59'5.00" N 14°28'5.00" E
Praha 4 - Libus	Residential area, road in the vicinity	Suburban	301	50°0'27.00" N 14°26'56.00" E
Praha 1 - nam. Republiky	The city centre, impact of the traffic	Urban	195	50°5'18.00" N 14°25'46.00" E

Moreover the ambient ozone concentrations have been correlated with meteorological parameters considered important for the ozone formation (global solar radiation, temperature and relative humidity).

Exposure Index AOT40

AOT40F value for forests was calculated using the following equation in accordance with Werner and Spranger eds. (1996):

$$AOT40F = \sum_{i \in V} \sum_{j=1}^{n} \sum_{k \in D} \left(c_{ijk} - p\right) \tag{1}$$

where

- c_{ijk} is ambient ozone concentration ≥ 40 ppb recorded in the i-th month, j-th day and k-th hour,
- p is a threshold value of 40 ppb,
- V is a set of months of the vegetation season,
- D is a set of daylight hours, defined as those hours with a mean global radiation of 50 $W.m^{-2}$ or greater,
- n is a number of days in a month.

AOT40 values for comparison with threshold level set for forests as 10 ppm.h were calculated for a six-month period beginning April 1.

For crops, as well as for semi-natural plant communities, the value of exposure index AOT40 C for comparison with critical level of 3 ppm.h was calculated either as a maximum running 3 month mean for the relevant vegetation period, or for a fixed period May-July (Kärenlampi and Skärby eds., 1996). We opted for calculation of the maximum running 3 month mean and used the following equation:

$$AOT40C = \max_{i \in W} \sum_{j=i+1}^{i+91} \sum_{k \in D} (c_{jk} - p), \tag{2}$$

where

- c_{jk} is ambient ozone concentration ≥ 40 ppb recorded in the j-th day and k-th hour,
- p is a threshold value of 40 ppb,
- j is a serial number of a day in vegetation period from April 1 to September 30,
- W is $<0; 92>$,
- D is a set of daylight hours, defined as those hours with a mean global radiation of 50 $W.m^{-2}$ or greater.

In 1994, when we started to calculate AOT40, there was a lack of global radiation measurements at the ambient ozone monitoring stations in the Czech Republic. The "daylight hours", defined as hours with a mean global radiation of 50 $W.m^{-2}$ or greater, as required by equations (1) and (2), could not, therefore, be strictly followed. Due to this fact the approach has been simplified and the "daily window" for ambient ozone concentrations relevant for exposure index calculation has been established for a fixed number of hours; i. e. 8–18 h UTC (Universal Time Co-ordinated) throughout the Czech Republic territory over the entire vegetation period (Hůnová and Novák, 2001).

Maps

The critical level exceedance fields for vegetation seasons during period 1994–2001 were derived from the AOT40 values calculated for monitoring sites. The results are presented in maps produced using the ARC/INFO system, version 7.2.1, module GRID.

For interpolation between the values at monitoring stations, the Inverse Distance Weighted Interpolation (IDW) method (Isaaks and Srivastava, 1989), was employed. The interpolated values have been calculated from the recorded values at the closest neighbouring stations according to the following equation:

$$Z(s_0) = \frac{\sum_{i=1}^{n} \frac{Z(s_i)}{h_{ij}^{\beta}}}{\sum_{i=1}^{n} \frac{1}{h_{ij}^{\beta}}} \qquad (3)$$

where

$Z(s_0)$	is the interpolated grid value,
$Z(s_i)$	is the neighbouring data point,
h_{ij}	is the distance between the grid node and the data point,
β	is the weighting power (the Power parameter),
n	is the number of measuring points.

The IDW method has been modified by combining it with weighting grid, which was orography. For each monitoring site a certain weight and spatial representativeness were specified. Only the concentrations recorded at monitoring stations outside the cities were taken into account in calculations. A radius of 10 km for urban background and a radius of 75 km for rural stations were allocated. Interpolation of values between stations was carried out in two stages. First the values were interpolated according to their weight and distance in a grid of 25 x 25 km, which made the network more detailed. From this network a 1x1 km grid in combination with the orography grid was calculated. For this presentation of the reported results, the 2x2 km grid, commonly used for construction of similar maps for other ambient air pollutants in the Czech Republic, has been selected.

The method takes into account the altitude´s effect on ozone concentration. This was introduced by the linear regression equation calculated from AOT40 values of rural and urban background monitoring sites.

The data from 19 monitoring sites for 1994 and 26 sites for 1997 have been used for deriving of maps; data coverage was taken into account in site selection.

RESULTS

Regression between Surface Ozone and Selected Meteorological Parameters

The surface ozone concentrations are strongly related with meteorology. Figs. 3–5 show statistically significant dependence of ozone concentrations (expressed as maximal daily 8-h means) on global radiation, temperature and relative humidity (mean 24-h values for each meteorological parameter considered). The ozone concentrations generally grow with increasing temperature and increasing global radiation and decrease with increasing relative humidity. In the case of global radiation and relative humidity the dependence is linear, while in the case of the temperature the dependence is better described as quadratic polynomial.

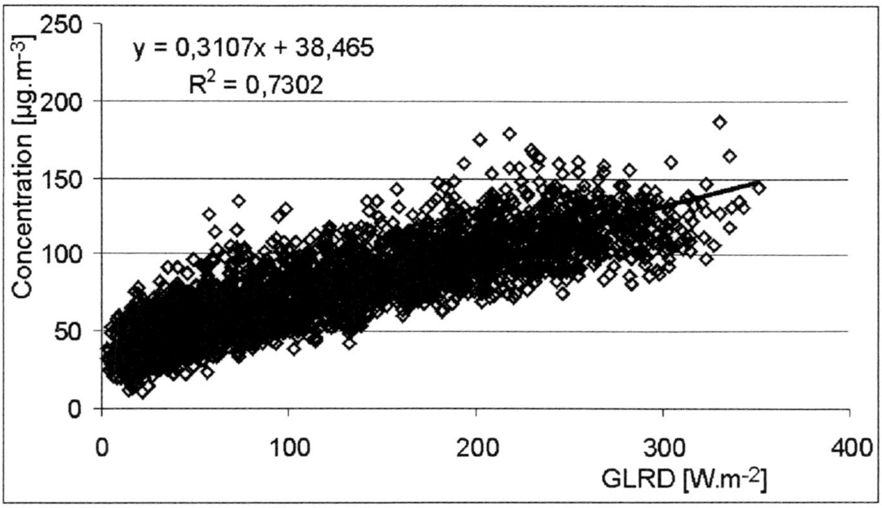

Fig. 3: The regression between ozone concentrations and global radiation

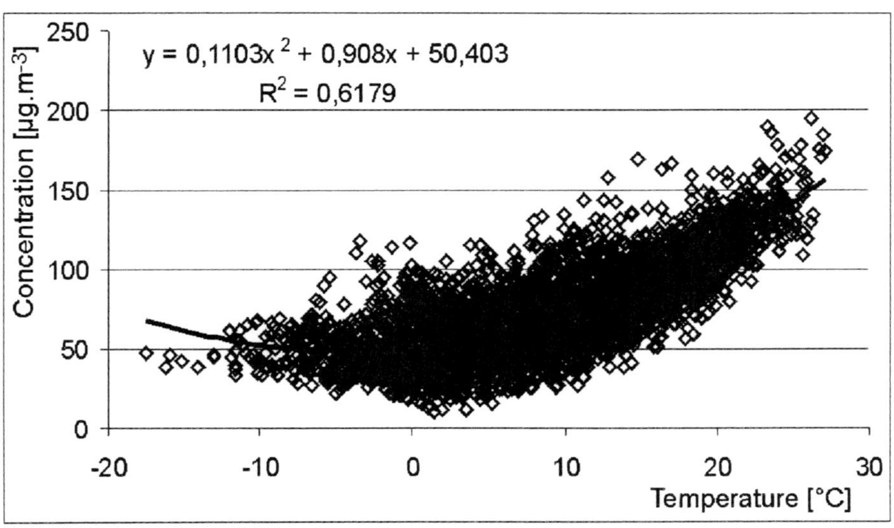

Fig. 4: The regression between ozone concentrations and temperature

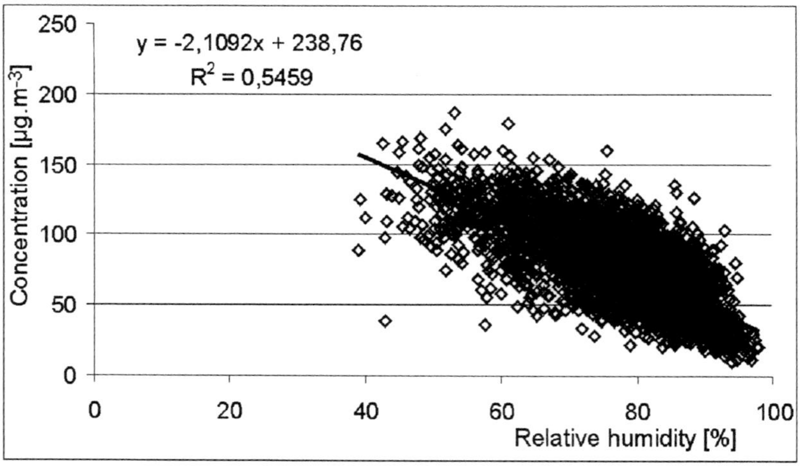

Fig. 5: The regression between ozone concentrations and relative humidity

Temporal Patterns

Annual Trends

The surface ozone concentrations depend strongly on meteorological conditions of the respective year (mostly summer period). The average 50th percentile, however, appears to stagnate, more or less, between 1994 and 2003 (Fig. 6), while the peak values expressed as 99.9th percentile and the 10th highest value decreases in contrast (Fig. 3). The decrease of peak levels during the period under review is most probably due to a certain reduction of precursor emissions. Also the thorough data check, not used formerly and introduced only recently, applied in order to identify the outliers (the method based on the special check of the differences of consecutive concentrations measured at one site combined with check of simultaneous concentrations at neighbouring sites), might add to the trend observed.

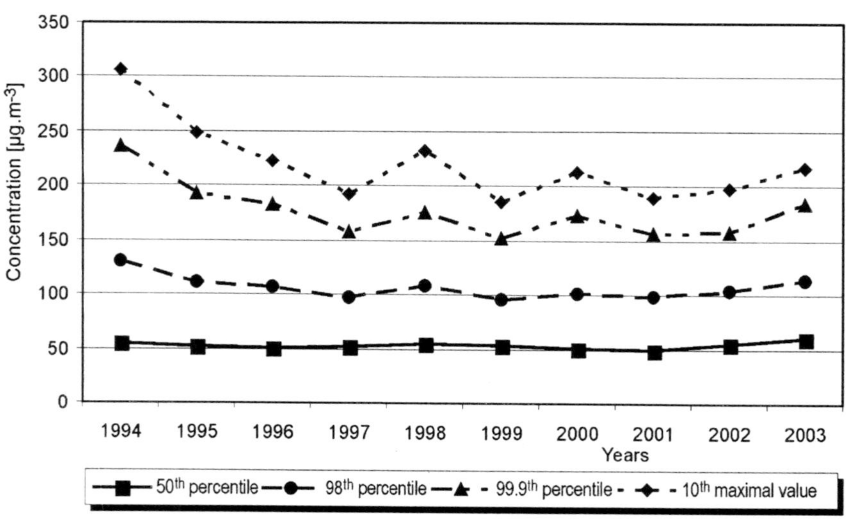

Fig. 6: The average 50th, 98th and 99,9th percentiles and the 10th highest value of surface ozone concentrations calculated from 12 stations measuring during the entire period of 1994–2003

As for the analysis of exceedances, it is evident that the last year 2003 – with extremely hot and sunny weather (high temperatures in the Czech Republic began as early as in spring and lasted until the autumn), provided conducive ozone conditions resulting in extensive episodes of high ozone concentrations. The year 2003 saw by far the highest number of limit value exceedances ever recorded in the Czech Republic. The threshold for health protection set up as the concentration of 120 $\mu g.m^{-3}$ in 8-h average was exceeded in 70 days in 2003, comparing to exceedance of 20–30 days per year in the period of 1996–2002 (Fig. 7). The exceedance of the threshold for vegetation protection set up as mean 24-h concentration of 65 $\mu g.m^{-3}$ was even more dramatic. This level was exceeded in nearly 160 days (44 % of days per year) in 2003 (Fig. 7).

The data analysis carried out for stations grouped according to the types, shows similar trends both for urban and rural stations. They differ, however, in much higher levels of concentrations recorded at the rural sites, compared to the urban ones, resulting in higher number of exceedances accordingly.

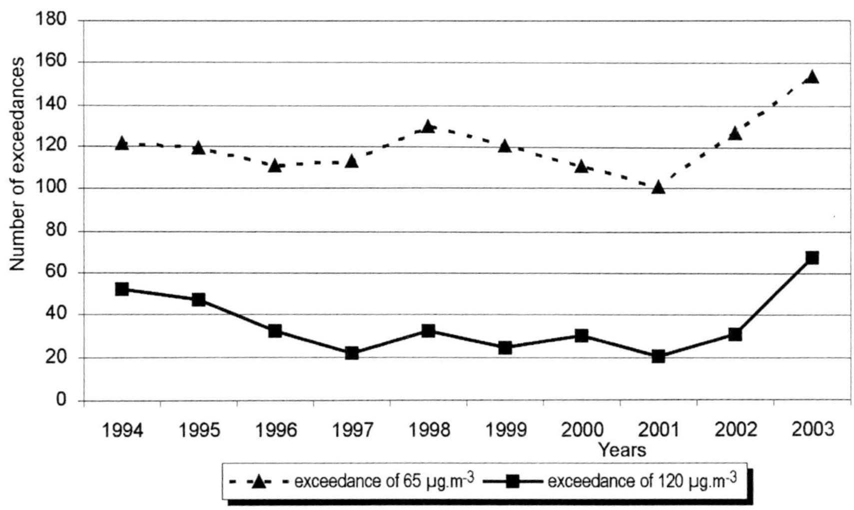

Fig. 7: The exceedances of threshold value for vegetation and ecosystems protection (24-h mean concentration of 65 µg.m–3) and limit value for human health protection (the maximal 8-h running mean concentration of 120 µg.m–3)

Seasonal Variability

The surface ozone is a typical summer pollutant in the Czech Republic. Fig. 8 reveals the seasonal variability at some stations in 2003, year rich in ozone.

Table 2 summarises the basic statistics for ozone concentrations for mountain stations Sous, Krkonose-Rychory, Churanov, Hojna Voda and Rudolice v Horach. Average monthly concentrations and monthly concentration ranges for the vegetation seasons of 1994–2003 are presented. The longest ozone data record is at site Sous, other sites started to record surface ozone later. The missing values in mean concentrations column throughout the Tab. 2 indicate that there was the insufficient data coverage in respective months.

Table 2: 24-h mean, maximal and minimal concentrations (μg.m^{-3}) for summer season months, monitoring sites Sous, Krkonose-Rychory, Rudolice, Churanov and Hojna Voda, 1994-2003

Hojná Voda

Months	1994		1995		1996		1997		1998		1999		2000		2001		2002		2003	
	Mea n	Min Max	Mea n	Min Max	Mean	Min Max	Mean	Min Max	Mean	Min Max	Mean	Min Max	Mea n	Min Max	Mea n	Min Max	Mean	Min Max	Mea n	Min Max
April			61.9	31.9 98.5		57.2 92.8		65.9 131.6	91.7	72.1 114.9	86.4	51.8 118	91.2	55.1 130.3	65.7	42.9 93.2	81.1	34.2 113.7	93.4	62.1 131.6
May			78.8	55.1 121.7	78.5	30.4 134.5	96.4	59 132.4		78.9 143.6	95.1	62.8 128.9	103. 3	73.6 138.2		34.2 90.9	88.1	40.9 125.7	91.4	57.1 125.7
June			57.8	32.9 94	96.3	57.5 130.9	93.6	73.6 125.7		63 121.6	88.8	58.9 123.8	105. 3	71.3 152.4	86.2	57.8 127.2	89.1	54.5 121.3	103. 0	58.6 136.8
July			82.6	44 125.1	80.4	59.5 120.2	78.8	50 109.7	82.3	65.4 123.7	92.5	48.8 131.5	78.3	53.9 120.9	84.9	40.9 135	86.1	44.5 123.4	96.8	69 143.6
August		48.0 69.4	69.2	43.3 94.3	84.3	51.5 114.4	90.0	40.4 129.8	101.7	56.5 149.8	87.8	48.2 138.3	96.4	60.5 137.9	88.3	51.3 131.8	86.9	53.5 121.1	110. 4	56.1 152.5
Sept.	51.6	29.5 75.9	51.0	33.7 80.1	53.0	27.4 77.2	76.4	44.1 115.2	67.8	49 97.6	86.8	57.9 125.7	64.6	24.3 96.5	61.4	45.9 84.1	65.4	30.7 112.5		58 112.9

Churáňov

Months	1994		1995		1996		1997		1998		1999		2000		2001		2002		2003	
	Mean	Min Max	Mea n	Min Max	Mean	Min Max	Mean	Min Max	Mean	Min Max	Mean	Min Max	Mea n	Min Max	Mean	Min Max	Mean	Min Max	Mean	Min Max
April					110.7	66.1 167.1	92.1	66.5 142.9	95.2	58.9 118.8	101.1	68.3 143.2	95.9	53.4 130.8	87.6	60.3 122.4	92.6	47.8 127.1	104.0	66.1 141.8
May				88 154.9	87.0	36.2 133	98.8	59.6 140.8	108.1	73.3 149	97.5	76 127.9	104. 4	73.7 141.6	95.8	38.6 141	94.0	56 149.6	96.6	63.8 135.9
June			77.0	42.7 127.2	98.3	63.2 146.8	94.5	67.2 133.3	101.5	69.3 142.1	87.8	60.1 131.4	101. 6	63.9 148.8	93.3	62.2 148.3	99.9	70.4 135.4	116.3	82.6 145.9
July			110. 8	59.1 158	85.0	59.6 119.5	85.0	55.7 112.7	88.2	67 134.6	91.3	56.6 131.2	77.7	53.4 121.4	91.3	45 143.3	91.6	50.2 135.8	93.8	64.5 134.7
August			100. 0	61 138.4	88.1	63.2 127.6	102.1	54.8 146.8	106.9	61.1 170.3	86.9	56.7 133.4	96.1	58.8 133.7	89.4	55.9 129.8	93.2	48.3 132.3	121.7	60.8 185.4
Sept.			68.0	45.8 92.3	58.0	26.8 86.7	85.9	56.6 116.5	77.5	58.4 106.5	86.1	59.4 129.2	70.5	18.5 119	62.5	50.9 86	73.4	44.2 113	94.9	57 161.5

Table 2: Continued

Krkonoše-Rýchory

Months	1994		1995		1996		1997		1998		1999		2000		2001		2002		2003	
	Mean	Min Max	Mean	Min Max	Mean	Min Max	Mean	Min Max	Mean	Min Max	Mean	Min Max	Mean	Min Max	Mean	Min Max	Mean	Min Max	Mean	Min Max
April					138.2	88.4 205.8	86.1	70.4 100.8	91.5	68.4 120.5	102.8	56 140.5	105.5	79.1 136.1	95.2	71.4 126.5	96.5	63.6 126.4	108.2	76.4 147.9
May					109.1	54.4 158.1	88.8	68.2 111.9	107.7	75.4 145.2		108.9 142.8	109.0	79 162.9	93.2	38.8 126.4	104.0	72.4 126.8	112.3	75.6 153.4
June					126.1	79.8 171.4		80.1 115.4	99.6	54.6 137.9	96.9	74 122.3	99.5	48.6 158.6		51.6 96.2	100.2	56.9 136.5	123.6	87.6 153.3
July					53.0	32.1 71		80.9 93.2	91.1	53.2 138.1	95.7	59.7 129.7	56.9	24.6 102.6	75.1	52.4 114.1	105.1	68.6 149.1	109.8	75.8 153.1
August			60.3	30.8 86.1	57.2	38.1 85.7		68.9 129	100.9	59 166.6	95.0	62.5 133.8	96.5	48 148.8	85.2	54.3 148.7	118.8	73.8 168.6	120.6	80.4 178.7
Sept.				26.9 62	40.8	22.5 82.2	89.1	70.4 114.2	76.3	52.9 99.3	95.4	58.7 143.5	74.8	44.3 112.8	66.8	37.1 89.1	89.6	56.3 137.7	99.5	63.4 168.8

Rudolice v Horách

Months	1994		1995		1996		1997		1998		1999		2000		2001		2002		2003	
	Mean	Min Max	Mean	Min Max	Mean	Min Max	Mean	Min Max	Mean	Min Max	Mean	Min Max	Mean	Min Max	Mean	Min Max	Mean	Min Max	Mean	Min Max
April					99.2	59.4 141.1	63.0	45.3 109	82.6	63 100.7		70.5 133.5	90.8	61.4 125.4	86.4	65.6 110.4	86.9	38.7 115.5	91.9	56.5 128.3
May					80.8	26.1 138.4	94.4	54.9 134	89.8	62.6 130.7	111.5	65.8 147.5	104.2	78.9 143	91.7	48.4 129	88.2	45.7 124.7	93.9	57 137.3
June					75.4	35.5 118.2	82.7	58.7 112.4	88.9	61.4 130	94.2	63.4 123.6	102.5	64.9 163.3	90.8	64 145	91.9	51.5 121.6	110.9	75.9 140.6
July					80.4	45 120.6	72.4	47.2 100.2	80.7	51.6 132.8	88.9	51.2 122.6	77.5	51.3 117.1	91.6	58.2 136.3	87.3	54 138.2	96.9	66.4 131.6
August					81.6	56.5 123.5	81.3	44 118.7	90.9	61.3 154.3	86.7	54.4 125.8	96.0	58.6 150.8	89.4	59.1 145.1	94.6	43.8 119.3	108.5	64 164.7
Sept.					52.4	25 75.5	61.8	17.4 101	67.0	47.8 81.8	89.1	56.4 132	66.2	43.6 97.4	58.2	36 89.1	72.5	42.1 107.3	87.6	52.7 156.6

Table 2: Continued

Souš

Months	1994		1995		1996		1997		1998		1999		2000		2001		2002		2003	
	Mean	Min Max	Mean	Min Max	Mean	Min Max	Mean	Min Max	Mean	Min Max	Mean	Min Max	Mean	Min Max	Mean	Min Max	Mean	Min Max	Mean	Min Max
April	93.6	51.3 133.1	87.6	55.3 136.6	110.6	69.6 158.6	80.5	58 99.5	93.7	64.6 129.1	84.2	47.3 131.1	85.4	58 118.4	77.2	54.4 112.2	83.9	39.6 123.2	90.3	57.7 133.2
May		67.7 131.9	94.8	61.8 149.4	77.4	41.2 106.4	87.5	60.4 115.5	95.0	63.9 136.6	92.1	61.1 112.9	91.4	66.1 134.6	83.5	37 119.5	87.4	31.7 123.5	88.9	45.8 127.9
June	100.4	57.2 169	78.5	56.4 98.1	74.8	42.2 105.4	89.6	60.8 117.5	89.9	60.9 132.9	81.2	57.2 111.8	90.8	56.3 140.3	76.3	54.3 117.7	80.8	56.4 114.5	98.7	74.9 136.4
July	133.2	92.6 174.6	102.2	67.4 135.5	59.2	34.4 82.4	78.8	53.5 104	82.5	48.8 132.3	82.3	57.8 113.9	63.7	44.8 87.4	79.0	50.8 126.2	81.0	50 134.6	88.3	65.5 128.9
August		66.2 162.7	95.8	58.5 136.9	78.1	44.1 123.8	94.6	56.1 139.2	88.7	59.4 146.9	74.1	40.6 112.8	76.6	40.5 103	72.5	42.6 113.4	88.0	63.1 117.8	93.8	67.2 150.6
Sept.	73.6	44.9 103.6	62.1	37.5 101.3		50 87.3	68.5	51.4 95.6	67.1	44.5 98.8	69.9	39.8 116.3	60.7	40.4 109.9	51.3	33.1 78.4	71.1	43.2 108.1	71.9	47.1 113

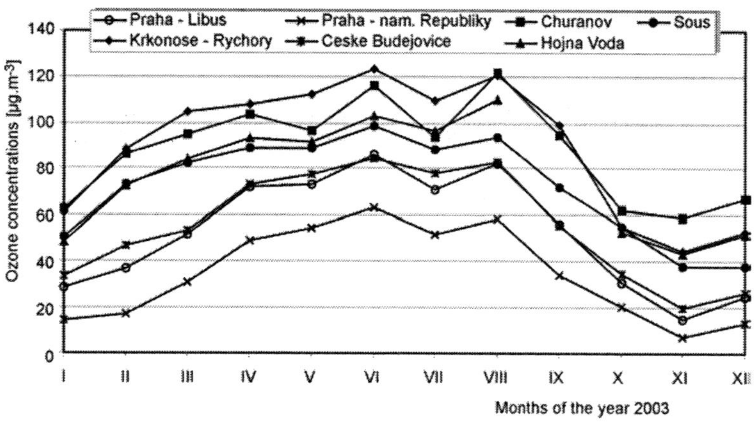

Fig. 8: Seasonal variability of ozone concentrations measured at different type of stations in 2003

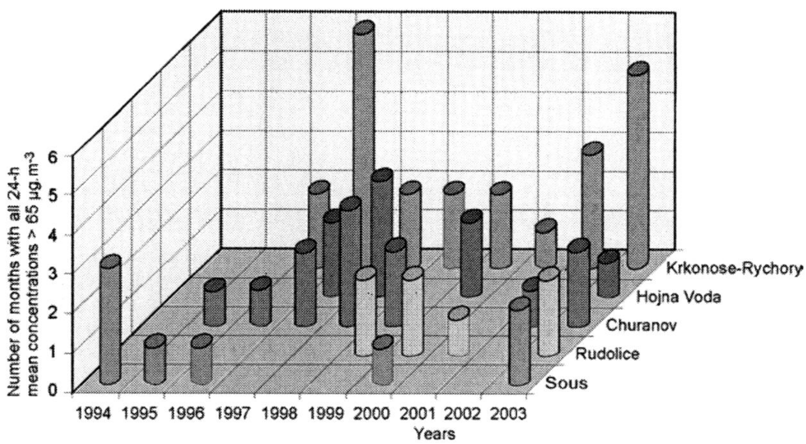

Fig. 9: The number of months with all 24-h mean concentrations higher than 65 µg.m-3 in vegetation seasons of 1994–2003 period

Exceedance of 24-h mean concentration of 65 $\mu g.m^{-3}$ at the Czech measuring sites is rather frequent. Fig. 9 presents the everyday exceedance of this threshold at the respective stations. There are many months at the mountains sites having all the 24-h mean concentration higher than 65 $\mu g.m^{-3}$.

Fig. 10 compares the vegetation seasons of 1994 and 2003, the years with extremely hot and sunny weather resulting in the highest surface ozone concentrations ever observed in the Czech Republic since the beginning of the O_3 monitoring. The comparison is made for the monitoring site Sous, which is one of few stations having the best data coverage for 1994. It is evident that the mean 24-h concentrations at Sous were higher in April, June, July and September of 1994 comparing to the concentrations in respective months in 2003. In May and June 1994 the criterion for data coverage needed for calculating the mean was not fulfilled. The maximal 24-h mean concentrations were higher in May, June, July and August 1994, while they were lower in September 1994 and equal in April 1994 comparing to the concentrations in respective months of 2003. The minimal 24-h mean concentrations were higher in May and July 1994, while they were lower in April, June, August and September 1994 comparing to the concentrations in respective months in 2003.

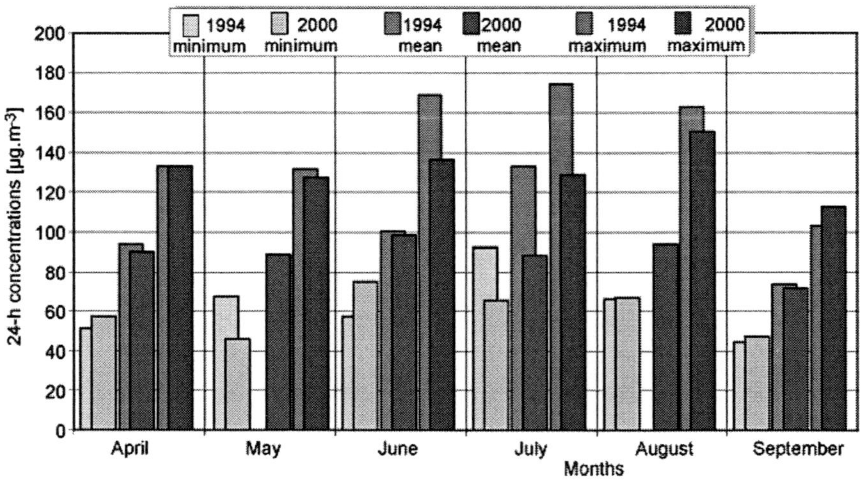

Fig. 10: The comparison of mean (a), maximal (b) and minimal (c) 24-h concentrations at the site Sous in vegetation periods of 1994 and 2003

Diurnal Variability

Fig. 11 presents a typical diurnal variability at different types of stations. In contrast to urban sites (represented here by the station Praha-Nam.Republiky, situated in the Prague centre close to the motor road) measuring significant ozone concentration decrease in the nighttime due to reactions with NO, the mountain sites (Churanov, Krkonose-Rychory) are characteristic by elevated ozone concentrations both during the daytime and night. The curve of diurnal variability at mountain sites tends to be much flatter than at the urban sites indicating the impact on elevated ozone concentrations on vegetation both during the day and night time. The rural sites, farther from the emission sources but not that high as the mountain sites, show the diurnal ozone concentration curve somewhere between urban and mountain sites. August 13, 2004, depicted here as an example, was a hot sunny day very suitable for ozone forming. The ozone concentrations reached the day maximum of 156 $\mu g.m^{-3}$ at the site Sous in Jizerske Mts. in the Northern part of the Czech Republic.

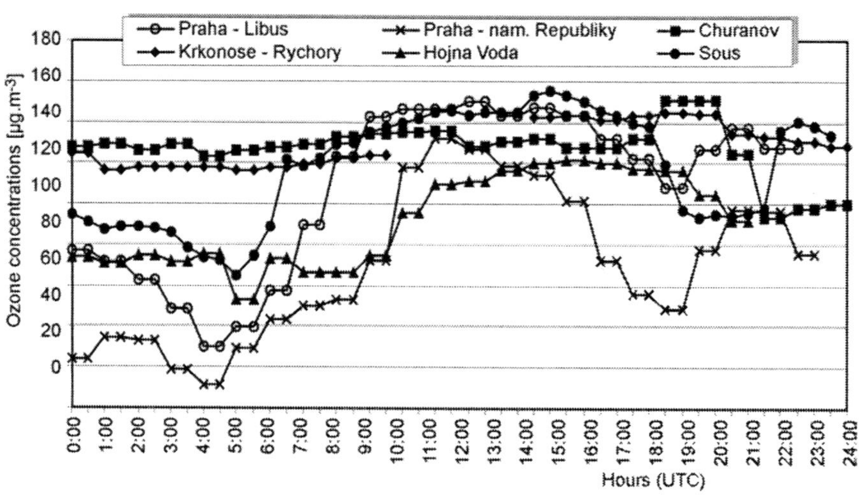

Fig. 11: Diurnal variability of ozone concentrations measured at different type of stations on August 13, 2004

Histograms

The comparison of histograms for surface ozone concentrations frequencies for the consecutive years shows similar pattern, both for urban and rural sites. The histograms for the year 2003 data show much higher occurrence of the higher concentration classes compared to the former years (see Fig. 12).

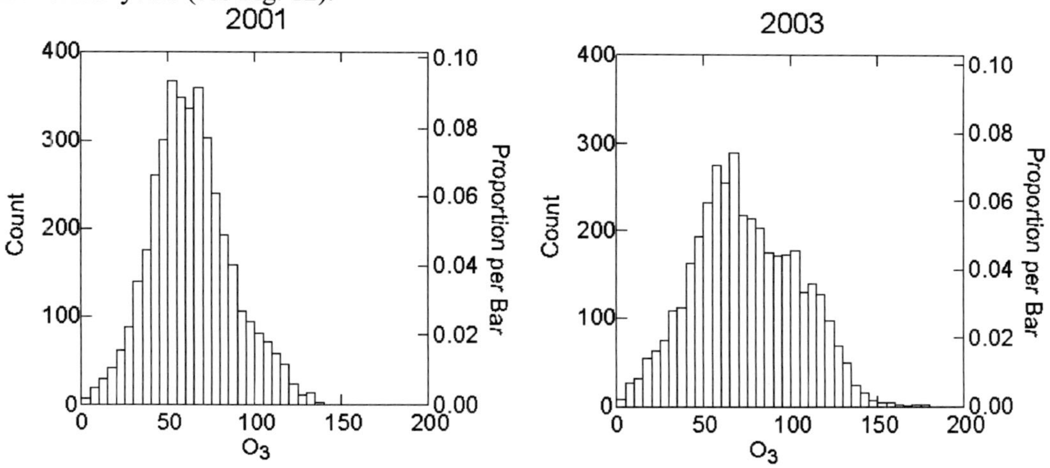

Fig. 12: The histograms showing the surface ozone concentration occurrences at the rural sites, for 2001 and 2003, averaged for 11 stations

Spatial Patterns

The AOT40 maps for forests and crops for 1994 and 1997 are presented at Figs. 13-16. Measuring sites the data of which was used as input data for calculation are indicated in maps. In addition to modelled AOT40 values, the total forested area (coniferous, broad leaved and mixed forests) and total agricultural area (arable land, permanent crops, pastures) have been depicted within Figures 13-16, respectively. Forested areas coincide generally with elevated ambient ozone concentrations as higher ozone concentrations are recorded in mountain areas where afforestation is high.

Table 3. Relative share of the entire territory of the Czech Republic for different exposure classes for forests (1994-2001) in %

AOT40 range [ppb.h]	1994	1995	1996	1997	1998	1999	2000	2001
0–10,000	0.09	2.29	0.87	14.90	3.40	3.30	0.14	5.50
10,000–15,000	0.76	3.63	37.75	43.80	31.36	32.67	16.58	73.98
15,000–20,000	9.60	33.61	59.44	40.35	58.91	61.07	67.09	20.46
20,000–25,000	22.49	46.22	1.91	0.96	6.23	2.88	16.19	0.06
25,000–30,000	53.89	8.48	0.04	0	0.11	0.07	0.01	0
30,000–35,000	12.81	5.66	0	0	0	0	0	0
> 35,000	0.36	0.12	0	0	0	0	0	0

Spatial critical level exceedances both for forests and crops/semi-natural vegetation are calculated for vegetation seasons of 1994 through 2001. The relative share of the entire territory of the Czech Republic for different exposure classes is presented in Tab. 3 (for forests) and Tab. 4 (for crops).

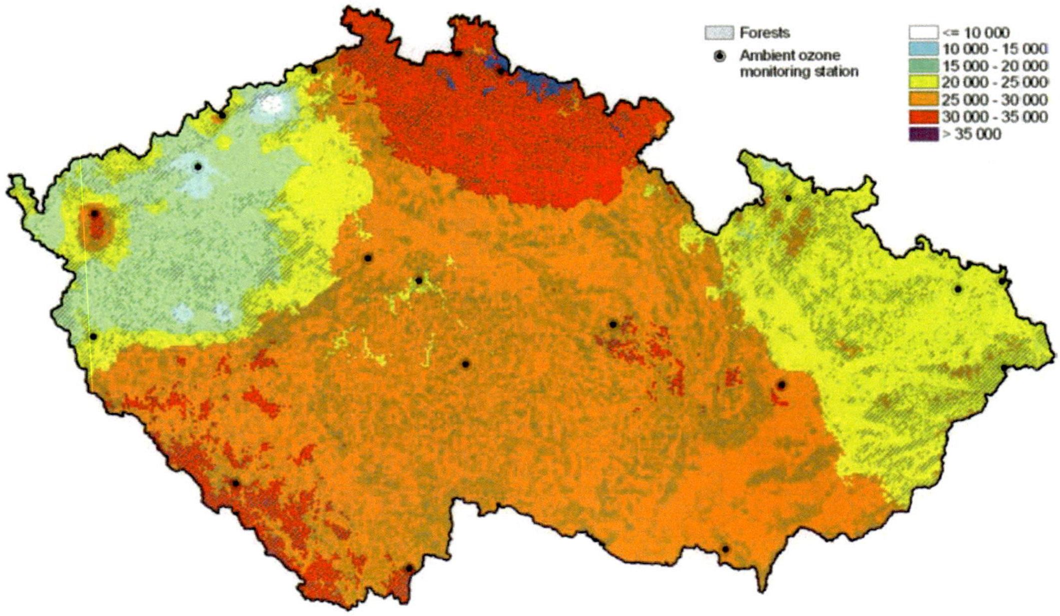

Fig. 13: Fields of AOT40 ambient ozone exposure index for forests [ppb.h], 1994

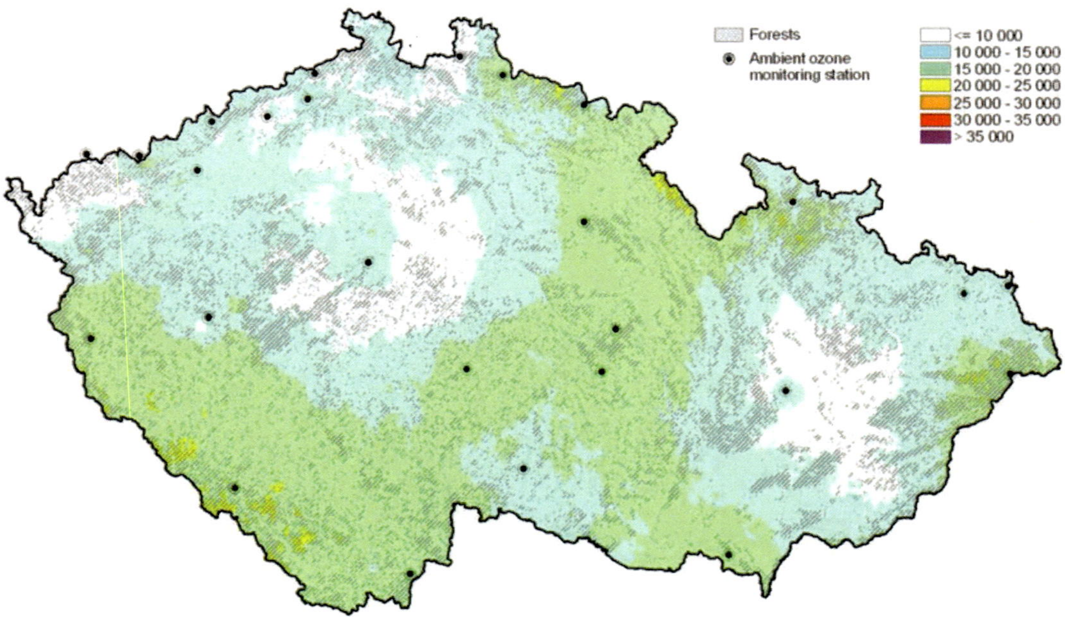

Fig. 14: Fields of AOT40 ambient ozone exposure index for forests [ppb.h], 1997

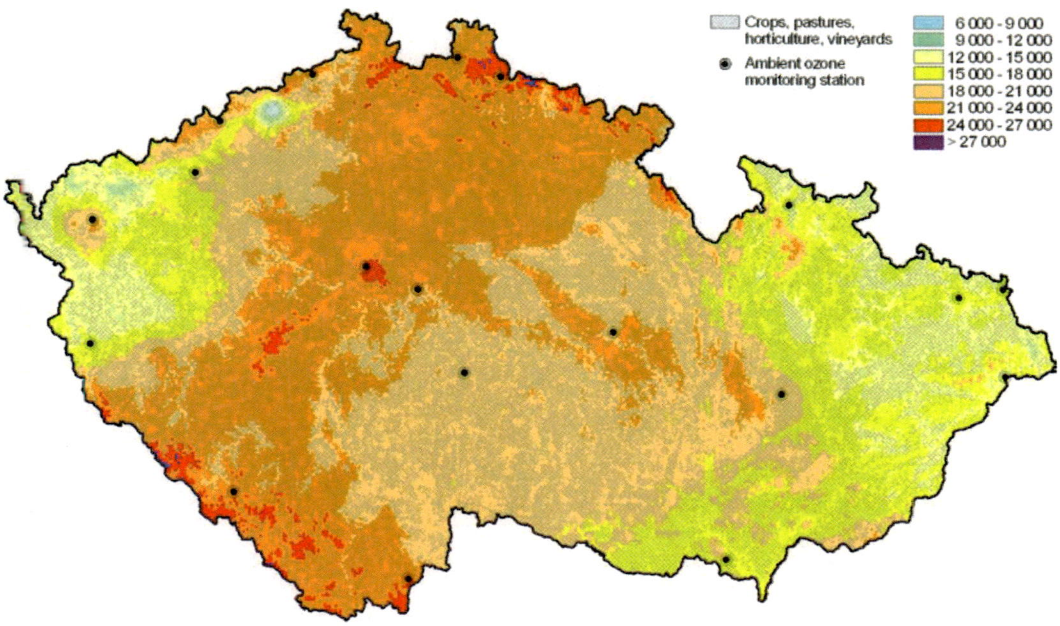

Fig. 15: Fields of AOT40 ambient ozone exposure index for crops and semi-natural vegetation [ppb.h], 1994

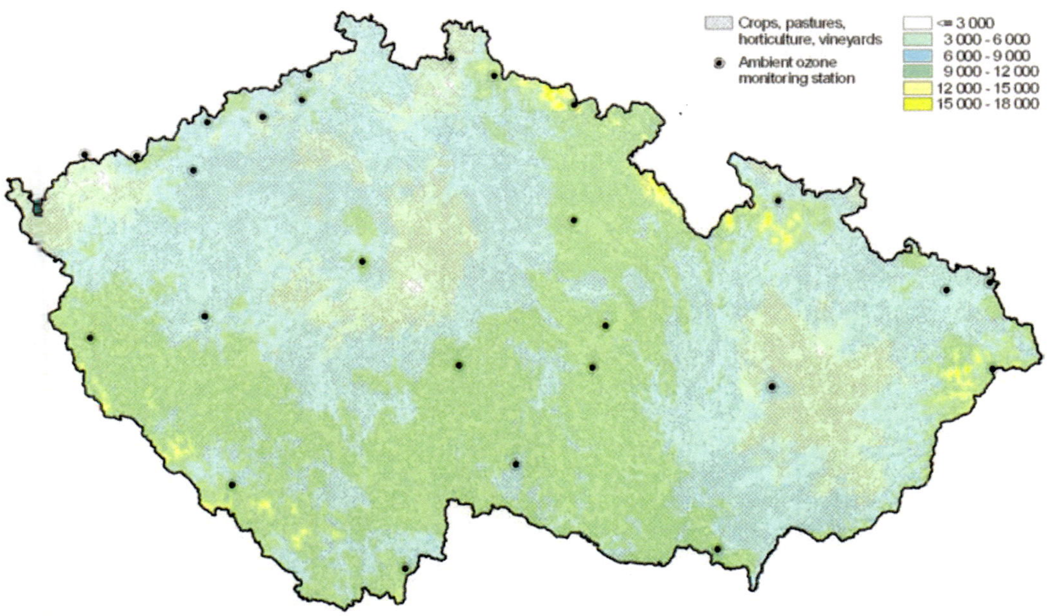

Fig. 16: Fields of AOT40 ambient ozone exposure index for crops and semi-natural vegetation [ppb.h], 1997

Table 4. Relative share of the entire territory of the Czech Republic for different exposure classes for crops (1994-2001) in %

AOT40 range [ppb.h]	1994	1995	1996	1997	1998	1999	2000	2001
0–3,000	0	0	0	0.23	0	0.02	0.03	0.05
3,000–6,000	0	0	0.56	14.27	0.99	3.21	0.25	4.18
6,000–9,000	0.04	0.04	11.95	45.87	25.33	35.90	6.82	60.71
9,000–12,000	0.24	6.38	53.03	38.18	58.89	57.38	49.19	33.86
12,000–15,000	10.05	48.38	29.49	1.33	13.93	3.31	41.79	1.18
15,000–18,000	18.42	25.94	3.00	0.11	0.83	0.19	1.79	0.02
18,000–21,000	36.55	18.05	1.61	0	0.03	0.001	0.14	0
21,000–24,000	32.77	1.16	0.31	0	0	0	0	0
24,000–27,000	1.85	0.06	0.06	0	0	0	0	0
> 27,000	0.08	0	0	0	0	0	0	0

The AOT40 threshold level of 10 ppm.h for forests was exceeded practically over the entire territory in 1994, 1996 and 2000 (more than 99 % of the territory over the threshold) while in other years 97.7 – 85.1 % of the forested territory was over the threshold (see Tab. 3). 1997 was the year with the worst meteorological conditions for ambient ozone formation during the reviewed period, and still 85.1 % of the Czech forested territory was above the threshold value of 10 ppm.h. In 1994 a very high share of AOT40 values were ranked in high exposure classes; 90 % of the Czech territory experienced the exposure values for forests between 20,000 and 35,000 ppb.h.

The AOT40 threshold level of 3 ppm.h for crops/semi-natural vegetation was exceeded over the entire territory of the Czech Republic for all intents and purposes during all years (see Tab. 4). Similarly as for forests, in 1994 a relatively high share of AOT40 values were ranked in high exposure classes; ca. 70 % of the Czech territory experienced AOT40 for crops/semi-natural vegetation between 18,000 and 24,000 ppb.h. In contrast, during 1997 through 2001 the exceedence of the 18,000 – 21,000 ppb.h levels did not occur.

The ambient ozone concentrations and consequently the cumulative exposure values expressed as AOT40 showed rather high year-to-year variability due to meteorological conditions during the vegetation period. In years with lower ozone concentrations (e. g. 1997) the threshold value for forests (10 ppm.h) was exceeded less than 2 times while in years with plentiful ozone (e. g. 1994, 1995) the threshold level over the most of the territory was exceeded 2–3.5 times. The threshold value for crops/semi-natural vegetation (3 ppm.h) in years with lower ozone concentrations was exceeded 5 times while in years with plentiful ozone it was exceeded as many as 8 times. Therefore a long-lasting high potential for harmful impacts of ambient ozone both for forests and crops/semi-natural vegetation was found for the entire territory of the Czech Republic.

DISCUSSION

The Input Data for the Assessment

The input data in the database is thoroughly checked before the data analysis. The method for detecting potential outliers is based on searching for differences between the measurements of the respective pollutant concentrations in two consecutive measurement terms. First the measured data are verified with regard to the occurrence of extremely high values and subsequently zero and too large differences. The methodology (Horálek et al., 2002), developed originally for 24-h data and used later for 30-min mean concentrations as well, is used routinely for checking the data in the Ambient Air Quality System in the Czech Hydrometeorological Institute.

This central database, gathering all the ambient air quality data – that is ambient pollutant concentrations, precipitation chemistry and atmospheric deposition and meteorological parameters needed for different ambient air quality calculations and interpretation of measured data – is a unique and excellent tool for the ambient air quality assessment.

The AOT40 Calculating and Mapping

Out of all 60 monitoring stations over the Czech Republic territory, only the sites located within rural and suburban areas were included in the calculations of AOT40; ozone values recorded at urban stations were considered not to be relevant for the AOT40 mapping. Considering the representativeness of the rural ambient ozone monitoring stations, a sufficient number of stations remained for producing AOT40 distribution maps. The sites used were selected carefully with respect to their data coverage as well. In addition to the Czech sites, the data from 2 German stations (Fichtelberg and Carlsfeld) located in the border area, were used for the 1997 analysis.

The number of stations monitoring ambient ozone in the Czech Republic is relatively high although there is a serious lack of stations in leeward areas of the big cities, where ozone concentrations are expected to be very high (EC, 1999). Thus the AOT40 values as presented within this paper might be underestimated in these areas.

The important problem when calculating AOT40 is determining the time period of any given day that is to be included. Due to the lack of solar global radiation measurements at the sites monitoring ambient ozone in the Czech Republic in the past, we estimated the "daylight window" for 8-18 hours UTC and this was applied uniformly for all stations and the entire vegetation period. When recently checking the differences between results of the above approach and the results obtained by strictly adhering to equations (1) and (2), the resulting differences were surprisingly low (Hůnová and Novák, 2001). The EC ozone directive establishes the "daylight window" for AOT40 calculation as 8-20 hours CET (Central European Time), i. e. 7-19 hours UTC. This caused a change in our hitherto calculation by broadening the "daily window" under consideration by 2 hours.

The IDW method used for point data interpolation is relatively uncomplicated and suitable when the statistical structure of the analysed field is not well known. The attribute of the method is a tendency to maintain the recorded values at the monitoring stations as much as possible. An inherent limitation of the method is an assumption that the influence of neighbouring points on interpolated values are exclusively a function of the distance between stations, which certainly is not true in the case of ambient ozone. Using another interpolation method, such as kriging, would require, however, more detailed analysis of the statistical structure of the data field analysed; such a more detailed analysis was not the aim of this study.

The authors believe that the introduction of altitude´s correction into AOT40 fields makes the maps closer to reality as the altitude´s effect on ambient ozone concentrations is significant. The O_3 (and thus the AOT40 values) generally increase with altitude (e.g. Gay, 1991; Brönniman et al., 2000). This does not have to be always a case as in mountain environment after exceeding 1200 – 1500 m a. s. l. decrease of ozone concentration with increasing altitude have been observed (Puxbaum et al., 1991; Bytnerowicz et al., 2002). Monitoring stations in the Czech Republic, however, are situated at altitudes up to 1122 m a. s. l. Most of our sites were under 1000 m a. s. l. with only 3 sites above 1000 m a. s. l.

The development of a meaningful ozone surface is a difficult task. Diem (2003) gives an excellent examination of ozone mapping techniques and draws attention to potential faults, often encountered in ozone mapping studies. He stresses the urgent need of spatial-scale approach lacking in many studies reviewed. Use of the different approaches to produce a map of ozone concentrations from monitored data to indicate problem areas is presented e.g. in the book on surface ozone in the Sierra Nevada (Fraczek et al., 2003; Lee, 2003; Preisler and Schilling, 2003).

We tried to make the analysis from all data available from the entire measuring period. The 1993 records are considered as the test run as it was the year when the monitoring of ambient ozone started, so this data is not used. The AOT40 is mapped until 2001 at present. The data is available and verified until the end of 2003. The AOT40 map construction is rather demanding, however, so currently the years 2002 and 2003 are not in their final version to be included in the assessment for this book.

The highest ozone concentrations and the most frequent and longest episodes since 1992 were observed in 1994 and 2003. The lowest concentrations and far less intensive ozone episodes were observed in 1997. It is worthwhile to mention that even in years with cold and rainy vegetation season (1997) the threshold levels for AOT40 were exceeded at large territory. Surprisingly, the site Krkonose-Rychory recorded O_3 concentrations above the vegetation protection threshold in all months of the vegetation season 1997 (see Fig. 9).

The ambient ozone data for 1999 in the Czech Republic causes further interpretations. The O_3 concentrations were below ambient air quality standards and there were no elevated concentration episodes in spite of relatively warm and sunny conditions throughout most of the vegetation period.

According to the annual ozone mapping (Hjellbrekke, 1998) based on data recorded at the EMEP stations (Cooperative Programme for Monitoring and Evaluation of the Long Range Transmission of Air Pollutants in Europe, referred to as the European Monitoring and Evaluation Programme), within the Czech Republic with portions of Austria and Switzerland there are areas where the recorded ambient ozone concentrations (expressed as exposure index AOT40) account for the highest readings in Europe. Current measurements demonstrate that ambient ozone concentrations substantially exceed the critical levels for crops, semi-natural vegetation and trees over large rural areas. This indicates that a potential risk of damages from ozone exist over most of Europe. The actual extent of ozone impacts over this area, however, is much less certain (Fuhrer et al., 1997).

Passive Samplers

In the Czech Republic relatively well-designed network monitoring, among other pollutants, concentrations of surface ozone, has been under operation since 1993. As already mentioned above, it consists of 60 stations covering evenly the entire territory, giving the 30-min concentrations measured by the UV-photometry (Hůnová et al., 2003). This network gives a good picture of overall ambient air quality though it is considered insufficient for the detailed assessment comparing the air quality and forest health status. In Purdon et al. (2004) the ambient air quality and forest health status in selected mountain regions in the Czech has been assessed on the basis of only one ambient air quality monitoring site for one mountain range. These sites are considered to be representative of the entire mountain range area. They do not allow, however, to provide more detailed analysis as the ozone concentrations vary due many terrain features as the elevation, slope exposure and vegetation cover. For more detailed assessment, a very useful method of passive samplers is available and draws more and more attention of the researchers.

The passive samplers are simple, inexpensive and their usage does not require the electricity power source. These advantageous properties make them a useful and continuously more popular method for ambient air quality measurement. The disadvantages of the method is the fact that we obtain integrated values for several days – several weeks and the sensitivity to meteorological conditions. The careful design and resulting reduction of the influence of meteorology on the sample can reduce the uncertainty substantially (Hofschreuder et al., 1999).

The interest in use of passive samplers for quantifying ambient, gaseous air pollutant concentrations, particularly in remote and wilderness areas is evidently growing among the scientists studying the links of air pollution and vegetation effects (Krupa and Legge, 2000). The passive samplers are used e.g. for ozone measurements in the U.S. National Parks where they have been shown to be useful in extending the continuous monitoring networks and for exploratory studies (Ray, 2001). The development of low cost robust passive monitors has enabled affordable landscape studies of air pollution patterns in remote areas. The study in the U.S. Sierra Nevada was among the first combining the new passive samplers with existing continuous monitors, and foliar surveys to study relationship between ozone and tree health status (Arbaugh and Bytnerowitz, 2003). The European Union also strongly supports the use of passive samplers for ambient air quality measurements. In Council Directive 96/62 EC on Air Quality and Management, a framework is set for preliminary assessment of air quality, station setting, supporting generalisation measurements and evaluation of existing measurements. In the Guidance report on preliminary assessment under EC directives (van Aalst et al., 1998) it is concluded that the low cost and easy operation of diffusive sampling techniques make them an ideal tool for large scale air pollution surveys with a high spatial resolution. The diffusive samplers are also of particular interest as an indicative technique.

For more detailed information on spatial distribution of ozone concentrations, the passive samplers have been applied also in the Czech Republic recently. The Willem´s badges for 24-h mean concentrations are used to record ambient ozone levels in selected mountain forested areas. The analytical method used is based on oxidation of blue indigosulfonate (IDS) by ozone to colourless isatinsulphonate, the concentration is derived from absorption measured by colorimetry. The students of the Institute for Environmental Studies at the Faculty of

Science, Charles′ University in Prague have already several years participated in the passive sampler monitoring within their diploma and PhD theses.

Due to the excellent references from other projects, the Ogawa samplers have been bought by the Czech Hydrometeorological Institute and they are currently under the test run in the Novohradske hory Mts. (Hůnová and Běláková, 2004), Orlicke hory Mts. and the National Park Ceske Svycarsko in the former "Black Triangle" region (highly industrialized and formerly highly polluted border zone area between southern Saxony in Germany, Silesia in SW Poland and northern Czech Republic). The Ogawa samplers are used for recording 2-week mean concentrations and preliminary results are very promising.

The ozone Ogawa sampler uses a filter coated with a nitrite-based solution. 2 parallel filters are mounted on both ends of one badge. The ambient ozone oxidises the nitrite to nitrate, which is after the exposure analysed by ion chromatography to determine the nitrate ion concentration used to calculate the total amount of ozone used in the chemical reaction. The Ogawa samplers are the most widely used passive ozone samplers. The Ogawa samplers are used widely in the U. S. (Arbaugh and Bytnerowitz, 2003). In Europe they were used on larger scale in the project on Carpathian Mountains (Bytnerowitz et al., 2004). Koutrakis et al. (1993) found that fluctuations in relative humidity (between 10 and 80 %) and temperature (between 0 °C and 40 °C) do not influence sampler performance at typical ambient O_3 levels (40–100 ppb).

Potential Impacts of Ozone on Vegetation

Considering the potential impacts of ozone on vegetation the knowledge of the responses of nation-wide species composition is of primary importance. Forested area roughly accounted for 33.4 % of the Czech Republic territory by the year 2000 (BFH, 2001). Forest tree species composition has remained highly stable over the last thirty years with the major share being spruce Picea abies (54 % in 2000), followed by pine Pinus sylvestris (17.6 % in 2000). The share of mixed oak Quercus spp. accounts for 6.3 % and European beech Fagus sylvatica for 6.0 %.

Considering the possible impact of ambient ozone on vegetation, the different sensitivity of species is of major importance. Plant sensitivity to O_3 varies widely both among species and individuals. It is determined both by genetics and environmental conditions. Comparing the species as to their sensitivity remains a critical issue, however. Over the past years a number of reviews has been published ranking the tree and plant species according to their sensitivity towards ozone exposure. Particular species are sometimes classified as very responsive while other times as resistant. This is due to the fact that comparison between different experiments is extremely difficult due to different experimental conditions. Moreover the parameters considered (visible injury, growth or physiological performance, biochemical changes, etc.), often vary.

In general, all exposure indices (AOT40, SUM0, SUM06, SUM08, W126) are based on ambient ozone concentrations. Recently, however, it has been proposed that stomatal flux of ozone would be more appropriate measure than exposure indices, as the ozone uptake by plant is what really counts and not the ambient concentrations. This complicates the issue even more, as effective flux of ozone by plants is driven by many variables as phenological stage of plant development, irradiance, temperature, water availability (vapour pressure deficit and soil moisture deficit), differences between plant species in stomatal activity, etc. (Wieser et al., 2000; Baumgarten et al., 2000).

Models to estimate stomatal ozone uptake related to plant phenology and environmental conditions are developed (Emberson et al., 2000 a, b) and their need is stressed (Panek et al., 2002 a). Emberson et al., (2000 a) present differences in the spatial distribution between stomatal flux and the exposure index AOT40 modelled for different types of vegetation on a European scale and indicate important implications for risk analysis. If AOT40 had been used as a basis for assessing risk, the highest values would be found in Central Europe; in contrast, values of the flux indices in central Europe were either comparable to, or lower than, those in the other grid squares (Emberson et al., 2002).

CONCLUSIONS

The ambient air pollution, resulting mostly from energy production, belonged to the major environmental problems in former Czechoslovakia between 1960s and 1980s. As a result of profound political and economic changes in Central Europe in 1989, the air pollution (esp. the concentrations of SO_2 and SPM) was significantly reduced. Apart from fine particles, the photochemical smog, the indicator of which is the surface ozone, nevertheless, remains a hot issue. Big portions of the Czech Republic regularly exceed the limits and thresholds for the vegetation protection.

The decrease of the ambient ozone levels is, due to its complicated atmospheric chemistry, an extremely challenging task. The countermeasures needed are mostly on regional or national level.

Due to the known negative impacts of elevated ambient ozone concentrations on vegetation and ecosystems, the mapping of areas of potential risk within the country is of primary importance. Though there is a reliable and advanced ambient air quality monitoring network recording the ambient ozone concentrations over the Czech territory, it proves very useful to supplement the standard active monitoring with passive samplers in order to get more detailed picture of spatial ozone patterns especially in the mountain forested areas. These measurements can be used e.g. to analyze the differences in concentrations in different altitudes, differently exposed slopes, etc. and will contribute to develop more accurate maps of areas under potential risk due to elevated surface ozone concentrations.

Apart from more detailed ambient monitoring there is a strong need to improving the mapping techniques, using fully the potentialities of geographical information systems with inbuilt supporting parameters as meteorology and terrain (altitude, slope exposure).

Unless the more sophisticated flux approach is developed for practical application, the AOT40 exposure index constitutes to be a useful tool for annual assessment of increased ambient ozone impact on ecosystems, especially for large scale, landscape level, comparisons of potential risks. Using this indicator requires, however, an awareness of its inherent shortcomings and a certain caution in AOT40 values interpretation.

References

Alscher, R. G., and Wellburn, A. R., Eds. (1994). *Plant Responses to the Gaseous Environment*. London, Chapman and Hall.

Arbaugh, M. J. and Bytnerowitz, A. (2003a). Introduction to a regional passive ozone sampler network in the Sierra Nevada. In: A. Bytnerowicz, M. J. Arbaugh, and R. Alonso (Eds.), *Ozone Air Pollution in the Sierra Nevada: Distribution and Effects on Forests* (pp. 157–164). Amsterdam, Elsevier.

Arbaugh, M. J., and Bytnerowitz, A. (2003b). Ambient ozone patterns and effects over the Sierra Nevada: Synthesis and implications for future research. In: A. Bytnerowicz, M. J. Arbaugh, and R. Alonso (Eds.), *Ozone Air Pollution in the Sierra Nevada: Distribution and Effects on Forests* (pp. 249–261). Amsterdam, Elsevier.

Ashmore, M. R. (2003). Surface Ozone Effects on Vegetation. In: J. R. Holton, J. A. Curry, and J. A. Pyle (Eds.), *Encyclopedia of Atmospheric Sciences*. London, Academic Press, Elsevier Science, Ltd.

Ashmore, M. R., and Fuhrer, J. (2000). New Directions: Use and Abuse of the AOT40 Concept. *Atmospheric Environment, 34,* 1157-1158.

Ashmore, M. R., and Wilson, R. B., Eds., (1994). *Critical Levels for Air Pollutants in Europe*. London, DoE.

Barbo, D. N., Chappelka, A. H., Somers, G. L., Miller-Goodman, M. S., and Stolte, K. (1998). *New Phytologist, 138,* 653.

Baumgarten, M., Werner, H., Häberle, K.-H., Emberson, L. D., Fabian,P., and Matyssek, R. (2000). Seasonal ozone response of mature beech trees (Fagus sylvatica) at high altitude in the Bavarian forest (Germany) in comparison with young beech grown in the field and in phytotrons. *Environmental Pollution, 109,* 431-442.

BFH (2001). *Forest condition in Europe* (2001 Executive report). Geneva and Brussels, UN/ECE and EC.

Brasseur, G. (2003). An Integrated View of the Causes and Impacts of Atmospheric Changes. In: G. P. Brasseur, R. G. Prinn, and A. A. P. Pszenny (Eds.), *Atmospheric Chemistry in a Changing World* (The IGBP Series, pp. 207–230). Berlin, Springer-Verlag,

Brönniman, S., Schuepbach, E., Zanis, P., Buchmann, B., and Wanner, H. (2000). A climatology of regional background ozone at different elevations in Switzerland (1992-1998). *Atmospheric Environment, 34,* 5191-5198.

Bytnerowicz, A., Godzik, B., Grodzińska, K., Fraczek, W., Musselman, R., Manning, W., Badea, O., Popescu, F., and Fleischer, P. (2004): Ambient ozone in forests of the Central and Eastern European mountains. *Environmental Pollution, 130,* 5–16.

Bytnerowicz, A., Tausz, M., Alonso, R., Jones, D., Johnson, R,. and Grulke, N. (2002). Summer-time distribution of air pollutants in Sequoia National Park, California. *Environmental Pollution, 118,* 187-203.

Chappelka, A. H., and Chevone, B. I. (1992). In: A. S. Lefohn (Ed.), *Surface Level Ozone Exposures and Their Effects on Vegetation* (pp. 271-324). Chelsea, MI, Lewis Publ., Inc.

Chappelka, A. H., and Samuelson, I. J. (1998). Ambient ozone effects on forest trees of the eastern United States: a review. *New Phytologist, 13,* 91-108.

Chappelka, A., Somers, G., and Renfro, J. (1999). Visible ozone injury on forest trees in Great Smoky Mountains national Park, USA. *Water, Air and Soil Pollution, 116,* 255-260.

Coe, H., Gallagher, M. W., Choularton, T. W., and Dore,C. (1995). Canopy Scale Measurements of stomatal and Cuticular O_3 Uptake by Sitka Spruce. *Atmospheric Environment, 29,* pp. 1413-1423.

Colbeck, I., and Mackenzie, A. R. (1994). Air Pollution by Photochemical Oxidants. Amsterdam, Elsevier.

Crutzen, P. J. (1988). Tropospheric ozone: An overview. In: I. S. A. Isaaksen (Ed.), *Tropospheric ozone* (pp. 3–32). Norwell, MA, Reidel.

Diem, J. E. (2003). A critical examination of ozone mapping from a spatial-scale perspective. *Environmental Pollution, 125,* 369–383.

EC (1992). *Council Directive 92/72/EEC of 21 September 1992 on air pollution by ozone.* Official Journal of the European Communities, No. L 297/1.

EC (1999). *Ozone Position Paper. Office for Official Publications of the European Communities*, Luxembourg.

EC (2002). *Directive 2002/3/EC of the European Parliament and of the Council of 12 February 2002 relating to ozone in ambient air.* OJEC, No. L 67, pp. 14-30.

EC (2003). *Ozone-climate interactions* (Air pollution research report no. 81). Luxembourg, EC.

EEA (1998). *Europe's Environment: The Second Assessment.* Kidlington, Elsevier.

EEA (2003). *Europe's Environment:The Third Assessment.* Copenhage, EEA.

Emberson, L. D., Ashmore, M. R., Cambridge, H. M., Simpson, D., and Tuovinen, J.-P. (2000 a). Modelling stomatal ozone flux across Europe. *Environmental Pollution, 109,* 403-413.

Emberson, L. D., Ashmore, M. R., Murray, F., Kuylenstierna, J. C. I., Percy, K. E., Izuta, T., Zheng, Y., Shimizu, H., Sheu, B. H., Liu, C. P., Agrawal, M., Wahid, A., Abdel-Latif, N. M., Van Tienhoven, M., De Bauer, L. I., and Domingos, M. (2001). Impacts of air pollutants on vegetation in developing countries. *Water, Air and Soil Pollution, 130,* 107-118.

Emberson, L. D., Wieser, G., and Ashmore, M. R. (2000 b). Modelling of stomatal conductance and ozone flux of Norway spruce: comparison with field data. *Environmental Pollution, 109,* 393-402.

Fangmeier, A., Bender, J., Weigel, H.-J., and Jager, H.-J. (2002). Effects of pollutant mixtures. In: J. N. B. Bell,, and Treshow M. (Eds.), *Air Pollution and Plant Life.* Chichester, John Wiley and Sons.

Fiala, J., and Ostatnická, J., Eds. (1995-2003). *Znečištění ovzduší na území České republiky/Air Pollution in the Czech Republic* (bilingual: Czech/English). Praha, ČHMÚ.

Finnan, J. M., Burke, J. I. and Jones, M. B. (1997).An Evaluation of Indices that Describe the Impact of Ozone on the Yield of Spring Wheat (Triticum aestivum). *Atmospheric Environment, 31,* 2685-2693.

Fishmann, J. (1991). The global consequences of increasing tropospheric ozone concentrations. *Chemosphere, 22,* 685-695.

Fowler, D., Cape, J. N., Coyle, M., Flechard, Ch., Kuylenstierna, J., Hicks, K., Derwent, D., Johnson, C., and Stevenson, D. (1999). The global exposure of forests to air pollutants. *Water, Air and Soil Pollution, 116,* 5-32.

Fowler, D., Flechard, Ch., Cape, J. N., Storeton-West, R. L and Coyle, M. (2001). Measurements of ozone deposition to vegetation quantifying the flux, the stomatal and non-stomatal components. *Water, Air and Soil Pollution, 130,* 63-74.

Fraczek, W., Bytnerowitz, A., and Arbaugh, M. J. (2003). Use of geostatistics to estimate surface ozone patterns. In: A. Bytnerowicz, M. J. Arbaugh, and R. Alonso (Eds.), *Ozone Air Pollution in the Sierra Nevada: Distribution and Effects on Forests* (pp. 215–248). Amsterdam, Elsevier.

Fuhrer, J., and Achermann, B., Eds. (1994). *Critical Levels for Ozone* (UN-ECE workshop report). Liebefeld-Bern, FAC.

Fuhrer, J., and Achermann, B., Eds. (1999). *Critical Levels for Ozone – Level II* (Env. Documentation No. 115). Berne, SAEFL.

Fuhrer, J., Skarby, L., and Ashmore, M.R. (1997). Critical Levels for Ozone Effects on Vegetation in Europe. *Environmental Pollution, 97,* 91-106.

Gay, M. J. (1991). Meteorological and altitudinal influences on the concentration of ozone at Great Dun Fell. *Atmospheric Environment, 25A,* 1767-1779.

Gerosa, G., Cieslik, S., and Ballarin-Denti, A. (2003). Micrometeorological determination of time-integrated stomatal ozone fluxes over wheat: a case study in Northern Italy. *Atmospheric Environment, 3,* 777–788.

Grünhage, L., Krupa, S. V., Legge, A. H., and Jäger, H.-J. (2004). Ambient flux-based critical values of ozone for protecting vegetation: differing spatial scales and uncertainties in risk assessment. *Atmospheric Environment, 38,* 2433–2437.

Harrison, R. M., Ed. (1996). *Pollution: Causes, Effects and Control* (3rd edition). Cambridge, Royal Society of Chemistry.

Harrison, R. M., Ed. (1999). *Understanding our Environment. An Introduction to Environmental Chemistry and Pollution* (3rd edition). Cambridge, Royal Society of Chemistry.

Hill, A. C. (1971). Vegetation: a sink for atmospheric pollutants. *J. Air Pollut. Control Ass. 21,* 341-346.

Hjellbrekke, A.-G. (1998). *Ozone Measurements 1996* (EMEP/CCC- Report 3/98). Kjeller, NILU.

Hofschreuder, P., van der Meulen, W., Heeres, P., and Slanina, S. (1999). The influence of geometry and draught shields on the performane of passive samplers. *J. Environ, Monit. 1,* 143–147.

Horálek, J., Hrdá, J., Fiala, J., and Brabec, M. (2002). Detection of Gross Errors in Air Pollution Data. *Meteorologický časopis/Meteorological Journal. 5,* 3–8.

Hůnová, I., and Běláková, Š. (2004). Ground-level Ozone Passive Monitoring in the Novohradske hory Mts., Vysoka site, the Czech Republic. In: C. Zerefos (Ed.), *Ozone* (Proceedings of the XX Quadrennial Ozone Symposium, June 2004, Kos, Greece, pp. 879–880). Athens, University of Athens.

Hůnová, I., Livorová, H., and Ostatnická, J. (2003). Ambient air quality and deposition trends at rural stations in the Czech Republic during 1993–2001. *Ecological Indicators, 3,* 35–47.

Hůnová, I., Šantroch, J., and Ostatnická, J. (2004). Ambient air quality and deposition trends at rural stations in the Czech Republic during 1993–2001. *Atmospheric Environment, 38,* 887–898.

Hůnová, I., and Novák,V. (2001). Exposure index AOT40 as a tool for assessment of ground-level ozone impact on ecosystems /Role of daylight hours included in calculation for resulting exposition/. *Meteorologické zprávy, 54,* 65-72. In Czech.

Intergovernmental Panel on Climate Chance, IPCC (2001). *The Scientific Basis, Contribution of Working Group 1 to the Third Assessment Report.* Cambridge, United Kingdom and New York, NY, USA, Cambridge University Press.

Isaaks, E. H, and Srivastava, R. M., (1989). *An Introduction to Applied Geostatistics.* Oxford, Oxford University Press Inc.

Jenkin, M. E., and Clemitshaw, K. C. (2002). Ozone and other secondary photochemical pollutants. In: J. Austin, P. Brimblecombe and W. Sturges (Eds.), *Air Pollution Science for the 21st Century* (pp. 285–338). Kidlington, Elsevier Science Ltd.

Karnosky, D. F., Percy, K. E., Thakur, R. C., and Honrath, R. E. Jr. (2003). Air pollution and global change : A double challenge to forest ecosystems. In: D. F. Karnosky, K. E. Percy, A. H. Chappelka, C. Simpson and Pikkarainen (Eds.), *Air Pollution, Global Change and Forests in the New Millenium* (pp. 1–41). Amsterdam, Elsevier.

Kärenlampi, L., and Skärby, L., Eds. (1996). *Critical Levels for Ozone in Europe: Testing and Finalizing the Concepts.* Kuopio , University of Kuopio.

Kelly, J. M., Samuelson, L., Edwards, G., Hanson, P., Kelting, D., Mays, A., and Wullschleger, S. (1995). Are seedlings reasonable surrogates for trees? An analysis of ozone impacts on Quercus rubra. *Water, Air and Soil Pollution, 85,* 1317-1324.

Koutrakis, P., Wolfson, J. M., Bunyaviroch, A., Froehlich, S. E., Hirano, K., and Mulik J. (1993). Measurement of Ambient Ozone Using a Nitrate-Coated Filter. *Anal. Chem., 65,* 209–214.

Krupa, S. V., and Legge, A. H. (2000). Passive sampling of ambient, gaseous air pollutants: an assessment from an ecological perspective. *Environmental Pollution, 107,* 31–45.

Krupa, S. V., Nosal, M., Ferdinand, J. A., Stevenson, R. E., and Skelly, J. M. (2003). A multi-variate statistical model integrating passive sampler and meteorology data to predict the frequency distributions of hourly ambient ozone (O_3) concentrations. *Environmental Pollution, 124,* 173–178.

Laurilla, T., Tuovinen, J.-P., Tarvainen, V., and Simpson, D. (2004): Trends and scenarios of ground-level ozone concentrations in Finland. *Boreal Environment Research, 9,* 167–184.

Lee, E. H. (2003). Use of auxiliary data for spatial interpolation of surface ozone patterns. In: A. Bytnerowicz, M. J. Arbaugh, and R. Alonso (Eds.), *Ozone Air Pollution in the Sierra Nevada: Distribution and Effects on Forests* (pp. 165–194). Amsterdam, Elsevier.

Lefohn, A. S., Laurence, J. A. and Kohut, R. J., (1988). A Comparison of Indices that Describe the Relationship between Exposure to Ozone and Reduction in the Yield of Agricultural Crops. *Atmospheric Environment, 22,* 1229-1240.

Lissková, K. (1998). *Porovnání vybraných expozičních indexů využívaných k hodnocení vlivu troposférického ozonu na vegetaci /Comparison of different exposure indices for estimation of ambient ozone impacts on vegetation.* (In Czech. Diploma thesis). Praha, ÚŽP Př FUK.

Long, S. P., and Naidu, S. L. (2002). Effects of oxidants at the biochemical, cell and physiological levels, with particular reference to ozone. In: J. N. B. Bell and M. Treshow (Eds.), *Air Pollution and Plant Life.* Chichester, John Wiley and Sons.

Manning, W. J. (1993). *Proc. 86th Ann. Meet. Of Air and Waste Maneg. Assoc.* (93-WA80.01). Denver, Co.

Massman, W. J. (2004). Toward an ozone standard to protect vegetation based on effective dose: a review of deposition resistances and a possible metric. *Atmospheric Environment, 38,* 2323-2337.

Matyssek, R., Wieser, G., Nunn, A. J., Kozovits, A. R., Reiter, I. M., Heerdt, C., Winkler, J. B., Baumgarten, M., Häberle, K.-H., Grams, T. E. E., Werner, H., Fabian, P., and Havranek W. M. (2004). Comparison between AOT40 and ozone uptake in forest trees of different species, age and site conditions. *Atmospheric Environment, 38,* 2271–2281.

McLaughlin, S. B., and Kohut, R. J. (1992). The effects of atmospheric deposition and ozone on carbon allocation and associated physiological processes in Red Spruce. In: C. Eagar and M. B. Adams (Eds.), *Ecology and Decline of Red Spruce in the Eastern United States*. Berlin-Heidelberg, Springer-Verlag.

Mészáros, E. (1993). *Global and Regional Changes in Atmospheric Composition.* Boca Raton, Lewis Publishers.

Mohnen, V. A. (1992). Atmospheric Deposition and Pollutant Exposure of Eastern U. S. Forests. In: C. Eagar and M. B. Adams (Eds.), *Ecology and Decline of Red Spruce in the Eastern United States*. Berlin-Heidelberg, Springer-Verlag.

Musselman, R. C., and Minnick, T. J. (2000). Nocturnal stomatal conductance and ambient air quality standards for ozone. *Atmospheric Environment, 34,* 719-733.

Muzika, R. M., Guyette, R. P., Zielonka, T., and Liebhold, A. M. (2004). The influence of O_3, NO_2 and SO_2 on growth of Picea abies and Fagus sylvatica in the Carpathian Mountains. *Environmental Pollution, 130,* 65-71.

Panek, J. A., Baldocchi, D. D., and Goldstein, A. H., (2002 a). The Need for Spatially and Functionally Integrated Models of Ozone Deposition to Sierra Nevada Forests. In: A. Bytnerowicz, M. J. Arbaugh, and R. Alonso (Eds.), *Ozone Air Pollution in the Sierra Nevada: Distribution and Effects on Forests* (pp. 325–358). Amsterdam, Elsevier.

Panek, J. A., Kurpius, M. R., and Goldstein, A. H. (2002 b). An evaluation of ozone exposure metrics for a seasonally drought-stressed ponderosa pine ecosystem. *Environmental Pollution, 117,* 93-100.

Pasuthová, J., Ryšková, L., and Uhlířová H. (1987). Poškození lesních dřevin ozónem (Práce VÚLHM č. 70). Praha, VÚLHM.

Penkett, S. A., Law, K. S., Cox, T., and Kasibhatla, P. (2003). Atmospheric Photooxidants. In: G. P. Brasseur, R. G. Prinn, and A. A. P. Pszenny (Eds.), *Atmospheric Chemistry in a Changing World* (IGBP Series, pp. 73–124). Berlin, Springer-Verlag.

Percy, K. E., Legge, A. H. and Krupa, S. V. (2003). Tropospheric ozone: A continuing threat to global forests? In: D. F. Karnosky, K. E. Percy, A. H. Chappelka, C. Simpson, and Pikkarainen (Eds.), *Air Pollution, Global Change and Forests in the New Millenium* (pp. 85–118). Amsterdam, Elsevier.

PORG (1997). *Ozone in the UK* (4th report of the Photochemical Oxidants Group). London, Department of the Environment, Transport and the Regions.

Preisler, H. K., and Schilling, S. (2003). Use of nonparametric local regression to estimate surface ozone patterns over space and time. In: A. Bytnerowicz, M. J. Arbaugh, and R. Alonso (Eds.), *Ozone Air Pollution in the Sierra Nevada: Distribution and Effects on Forests* (pp. 195–214). Amsterdam, Elsevier.

Prinn, R. G. (2003). Ozone, Hydroxyl Radical, and Oxidative Capacity. In: H. D. Holland and K. K. Turekian (Eds.), *Treatise on Geochemistry* (Vol.4: The Atmosphere).

Purdon, M., Cienciala, E., Metelka, V., Beranová, J., Hůnová, I., and Černý, M. (2004). Regional variation in forest health under long-term air pollution mitigated by lithological conditions. *Forest Ecology and Management, 195,* 355-371.

Puxbaum, H., Gabler, K., Smidt, S., and Glattes, F., (1991). A one-year record of ozone profiles in an Alpine Valley (Zillertal/Tyrol, Austria, 600-2000 m a.s.l.). *Atmospheric Environment, 25A,* 1759-1765.

Ray, J. D. (2001). Spatial distribution of tropospheric ozone in national parks of California" interpretation of passive sampler data. In Proceedings of the International Symposium on Passive Sampling of Gaseous Air Pollutants in Ecological Effects Research. *The Scientific World, 1,* 483–497.

Rondón, A. (1993). Photoinduced deposition of ozone on the plant leaf cuticule. *Science*

Seinfeld, J. H. (1996): Atmospheric Chemistry and Physics of Air Pollution. John Wiley and Sons, New York.

Sillman, S. (1999). The relation between ozone, NOx and hydrocarbons in urban and polluted rural environments. *Atmospheric Environment, 33,* 1821–1846.

Sillman, S. (2002). The relation between ozone, NO_x and hydrocarbons in urban and polluted rural environments. In: J. Austin, P. Brimblecombe and W. Sturges (Eds.), *Air Pollution Science for the 21st Century* (pp. 339–385). Kidlington, Elsevier Science Ltd.

Singh, O. N. and Fabian, P. (2003). Atmospheric Ozone: A Millenium Issue. European geosciences Union. Special Publication Series. Vol. 1. Copernicus GmBH.

Skarby, L., Ro-Poulsen, H., Wellburn, F. A. M., and Sheppard, L. J. (1998). Impacts of ozone on forests: a European perspective. *New Phytologist, 139,* 109-122.

Sofiev, M., and Tuovinen, J.-P., (2001). Factors determining the robustness of AOT40 and other ozone exposure indices. *Atmospheric Environment, 35,* 3521-3528.

Staehelin, J., Thudium, J., Buehler, R., Volz-Thomas, A., and Graber, W., (1994). Trend in surface ozone concentrations at Arosa (Switzerland). *Atmospheric Environment, 28,* 75-87.

Stanners, D. and Bourdeau, P., Eds. (1995). *Europe's Environment: The Dobris Assessment.* Copenhagen, EEA.

Tuovinen, J.-P., (2000). Assessing Vegetation Exposure to Ozone: Properties of the AOT40 Index and Modifications by Deposition Modelling. *Environmental Pollution, 109,* 361-372.

UN-ECE (1988). *UN-ECE Critical Levels Workshop Report.* Bad Harzburg.

US EPA (1996). *Air Quality Criteria for Ozone and Other Photochemical Oxidants* (Voll. II. EPA-600/P-93/06bF). Research Triangle Park, NC, US EPA,.

Van Aalst, R., Edwards, L., Pulles, T., De Saeger, E., Tombrou, M., and Tønnesen, D. (1998). *Guidance report on preliminary assessment under EC air quality directives* (Technical report No.11). Copenhagen, EEA.

Warneck, P. (1988). *Chemistry of the Natural Atmosphere.* San Diego, Academic Press, Inc.

Werner, B., and Spranger, T., Eds. (1996). *Manual on Methodologies and Criteria for Mapping Critical Levels/Loads and Geographical Areas where they are Exceeded.* Berlin, UBA.

WHO (2000). *Air Quality Guidelines for Europe* (WHO regional Publications, European series, No. 91). Copenhagen.

Wieser, G., Häsler, R., Götz, B., Koch, W., and Havranek, W. M., (2000). Role of climate, crown position, tree age and altitude in calculated ozone flux into needles of Picea abies and Pinus cembra: a synthesis. *Environmental Pollution, 109,* 415-422.

WMO (2003). *Scientific assessment of ozone depletion: 2002* (Global Ozone Research and Monitoring Project, Report No.47). Geneva, WMO.

Zhang L., Brook J. R., and Vet R. (2002). On ozone dry deposition – with emphasis on non-stomatal uptake and wet canopies. *Atmospheric Environment, 36,* 4787-4799.

In: New Trends in Ecology Research
Editor: A. R. Burk, pp. 37-70
ISBN: 1-59454-379-8

Chapter 2

Basic Concepts Related to Uncertainty in Empirical Data Regulating Predictive Power of Mass-Balance Models

L. Håkanson and J. M. Malmaeus

Dept. of Earth Sciences, Villavägen 16, S-752 36 UPPSALA, Sweden

Abstract

This article is divided into three main sections. The first section deals with with uncertainties in empirical data, defining and discussing e.g., the highest r^2 of predictive models, the sampling formula and uncertainty in empirical data, empirically based highest r^2, highest reference r^2, CV and sampling period, CV versus mean values, mean versus maximum values and spurious correlations.

The second section includes issues concerning model testing, i.e., calibration and validation, sensitivity tests and uncertainty tests using Monte Carlo techniques.

The third section introduces some basic ideas concerning dynamic mass-balance models for lakes. Algorithms for water mixing and outflow are given and efficient ways to define deep water, surface water, erosion and accumulation areas are discussed. This section also discusses time and area resolution in dynamic lake models and the most efficient way of utilizing distributed empirical data.

Background and Aim

How does inherent variations and uncertainties in empirical data constrain approaches to prediction? This section addresses this question in discussing one of the most fundamental concepts in predictive modelling, the uncertainty in the empirical y-variable used to validate model predictions. So, the focus here is not on concentrations of X but rather on coefficients of variation (CV = SD/MV; SD = standard deviation, MV = mean value).

The first part of this section discusses a key question in predictive ecosystem modelling: How high is the highest possible predictive success for a given target variable y? It is evident that many factors are involved concerning, e.g., sampling (such as the number of samples), analysis (e.g., the method and precision used in determining y), temporal and spatial variations, model structure (e.g., which model variables x_i are included), the reliability of the model variables and the statistical methods used to define predictive success, e.g., the r^2-value (= the coefficient of determination; calculated from a regression between modelled values and empirical data of y; see Håkanson, 1999). The r^2-value will be used here as a standard criterion of predictive success since this is a widely used concept in aquatic modelling. If a model is validated, i.e., tested against an independent set of data, the achieved r^2-value will not just depend on the uncertainty in the empirical y-value (i.e., the uncertainty in the y-direction), but also on the structuring of the model (i.e., which processes and model variables are accounted for and the empirical uncertainty of the model variables; the uncertainty in the x-direction). This section will discuss three r^2-values, 1. the highest reference r^2 (r_r^2), 2. the empirically based highest r^2 (r_e^2) and 3. the achieved r^2 when modelled y-values are compared to empirical y-values (r^2).

To exemplify the concepts, this section will use empirical data from the River Danube (a site close to Regensburg). Since 1995 measurements of chlorophyll concentrations and the composition of different algal groups have been carried out using a method based on delayed fluorescence, DF (Friedrich et al., 1998; Bodemer, 1998; Håkanson et al., 2003) and those data will be used here to illustrate important concepts and principles related to predictive power.

Note that there exist great differences in the representativeness and reliability of all model variables. All water chemical variables vary with time and sampling location in a given aquatic system. Characteristic CV-values for many important variables for mass-balance models will be discussed in this section. CV-values are probably the most commonly expressed statistic in contexts of ecosystem modelling (see Whicker, 1997). It is not likely that other statistical measures of uncertainty (see Gilbert, 1987) will substantially change the general conclusions about empirical uncertainties for aquatic variables discussed in this chapter.

The Highest R^2 of Predictive Models

Fig. 1 illustrates some fundamental concepts related to "the highest r^2". The data in the figure should emanate from several sampling occasions from river or lake sites for defined time intervals (such as days, weeks or months). There are uncertainties for all such values for all water variables. The CV-value for within site variability is always related to very complex climatological, biological, chemical and physical conditions. It has been shown (see Håkanson, 1999) that it is often possible to define a characteristic CV-value for a given variable, like 0.25 for chlorophyll in lakes. By definition, CV is also largely independent of n, the number of data used to determine MV and SD, if n is large enough (n > 4). CV should also be independent on analytical method provided that the main difference between the methods can be expressed by a calculation constant ($y_1 = \text{const}\cdot y_2$).

The inherent uncertainty or variability associated with a given target variable is illustrated by the vertical uncertainty bands (error bars) in fig. 1. This uncertainty will evidently influence the result of regressions where the given variable is used either as x or y-variable (e.g., the r^2-value). If CV for y is large, one cannot expect a model to predict y well.

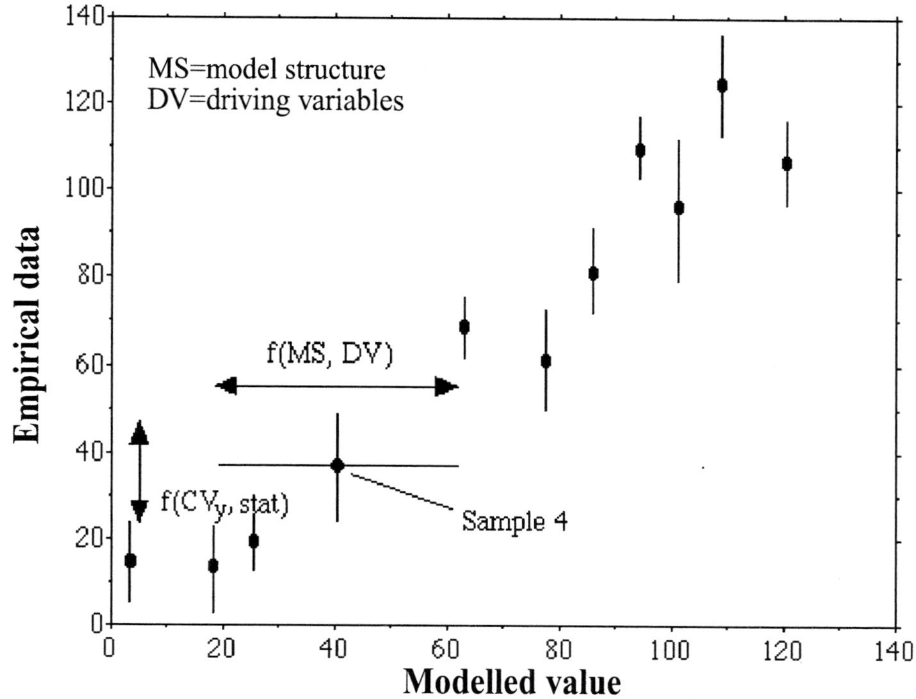

Figure 1. Illustration of fundamental concepts related to the question of "the highest possible r2" of models as determined from validations, i.e., a comparisons between modelled values (on the x-axis) and independent empirical data on the y-axis. Note that the uncertainty in the y-direction (e.g., given by the CV-value) differ among variables and that the characteristic uncertainty in the y-direction regulates the predictive power of the model.

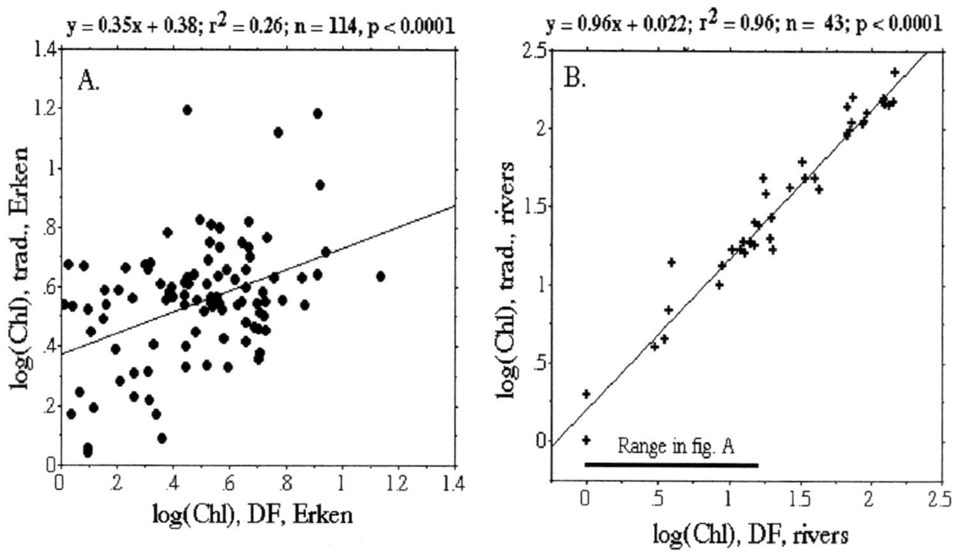

Figure 2. Two regressions between chlorophyll (logarithmic values) measured by the traditional extraction method (trad.) and with delayed fluorescence excitation spectroscopy (DF). A. Data from Lake Erken and B. Data from Rivers Danube and Naab (from Håkanson et al., 2003).

Fig. 1 illustrates a hypothetical model validation when modelled values of the target variable are put on the x-axis. The uncertainty in the x-direction is then related to the uncertainty associated with the model structure and the uncertainty of the model variables (x_i). Generally, one should expect the model uncertainty to be larger, or much larger, than the uncertainty in the y-direction. This is also illustrated in fig. 1.

The achieved r^2-value depends on the range of data. In fig. 2 two regressions are given between chlorophyll concentrations determined with two methods (either extracted chlorophyll or measurements using delayed fluorescence) for (A) Lake Ekoln, Sweden, and (B) river data to stress this methodological point. The r^2-values of the regressions are 0.26 and 0.96, respectively (using logarithmic values for chlorophyll). The mean difference [Diff = (M-E)/E; M = model regression value, E = empirical data] is, however, very similar in the two examples, 43% and 41%. The higher r^2-value in fig, 2B is mainly explained by the much wider range of data.

The Sampling Formula and Uncertainties in Empirical Data

If the within-site variability (CV_y or simply CV) is large, many samples must be at hand to obtain a given level of certainty in the empirical mean value. There is a general formula, derived from the basic definitions of the mean value, the standard deviation and the Student's t value, which expresses how many samples are required (n) in order to establish a mean value with a specified certainty (see fig. 3; see Håkanson, 1999).

$$n = (1.96 \cdot CV/L)^2 + 1 \qquad (1)$$

where L is the level of error accepted in the mean value. For example, L = 0.1 implies 10% error so that the measured mean will be expected to lie within 10% of the expected mean with the probability assumed in determining t. Since one often determines the mean value with 95% certainty ($p = 0.05$), the t-value is often 1.96 (or about 2).

The relationship between n, CV and L is illustrated in fig. 3. If the CV is 0.25 (the characteristic CV for chlorophyll in lakes), then 25 samples are required to establish a site-specific mean value provided that one accepts an error of L = 10%. If one accepts a larger error, e.g., L = 20%, fewer samples are required. This has profound implications for the quality of our knowledge of the variable. An important question, which will be addressed below is: How would CV depend on the length of the sampling period? Evidently, individual data from specific sites and sampling occasions will represent the characteristic/mean/median conditions very poorly when the CV-value is high.

Empirically Based Highest R^2, R_E^2

One way to determine the highest r^2 of a predictive model is to compare two empirical samples (r_e^2) since one can generally not expect models to predict better than empirical data. The data in these two samples should also be sampled, transported, stored and analysed in the same manner. To illustrate this, chlorophyll data from River Danube will be used. Fig. 4 shows that the r^2-value calculated from 178 samples is 0.77. This r^2-value indicates both analytical uncertainties and short-term temporal variability for chlorophyll in this river site. The more reactive a variable is, the more difficult it will be to establish representative empirical data of the given variable. Evidently, r_e^2 depends on the total number of samples in the regression, the number of samples for each individual value used in the regression, the standard deviations related to all individual data points, the standard deviation of all points in the regression and the range of the y-variable. So, the r_e^2-value is highly dependent on the sampling. The next r^2-value (r_r^2) has been defined as a standard reference r^2, which is basically not dependent on the sampling.

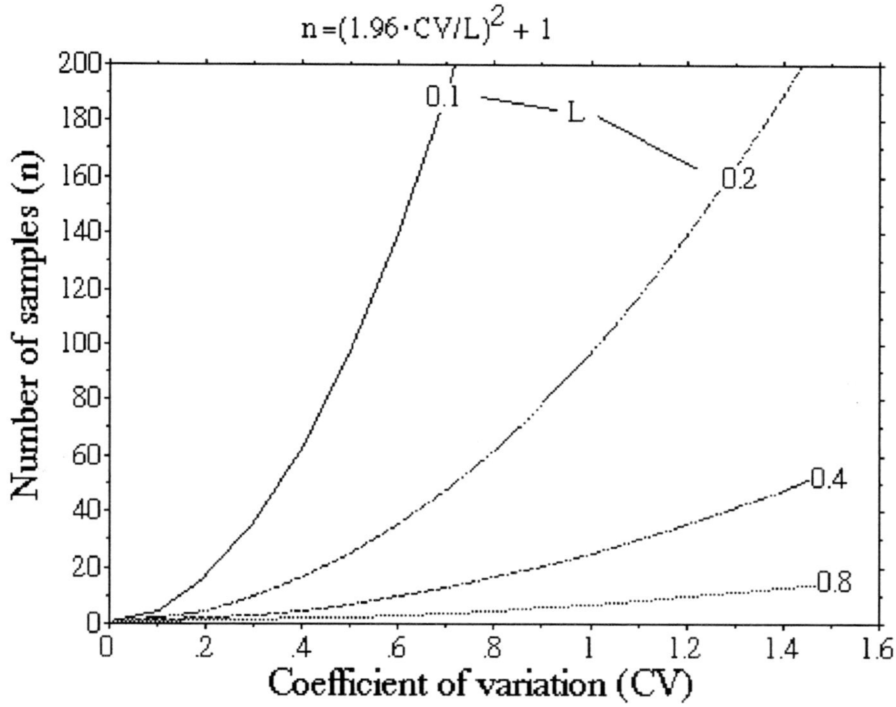

Figure 3. Illustration of the sampling formula. The nomogram shows how many samples that must be analysed (n) to establish a characteristic mean value (MV) with a given uncertainty or error (L) and a given confidence (95%).

HIGHEST REFERENCE r^2, r_r^2

The highest reference r^2, r_r^2, has been presented and motivated by Håkanson (1999). The r_r^2-value is defined as:

$$r_r^2 = 1 - 0.66 \cdot CV_y^2 \quad (2)$$

where CV_y is the characteristic within-site variability for the given y-variable . The equation is graphically shown in fig. 5. It is valid for actual (i.e., non-transformed) y-values. In practice, one cannot expect to obtain r^2-values higher than the r_r^2-values, so these values may be used as reference values for predictive models.

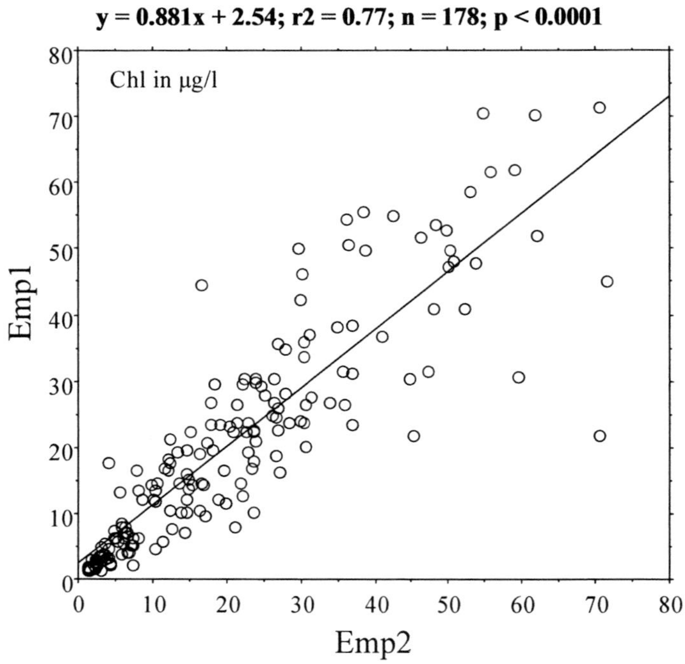

Figure 4. Determination of "empirically highest r2" (re2) for the chlorophyll-a concentration from a regression between two parallel empirical samples (Emp1 vs Emp2). The re2-value is 0.77 for the given 178 data-pairs for chlorophyll in the River Danube (data from Håkanson et al., 2003).

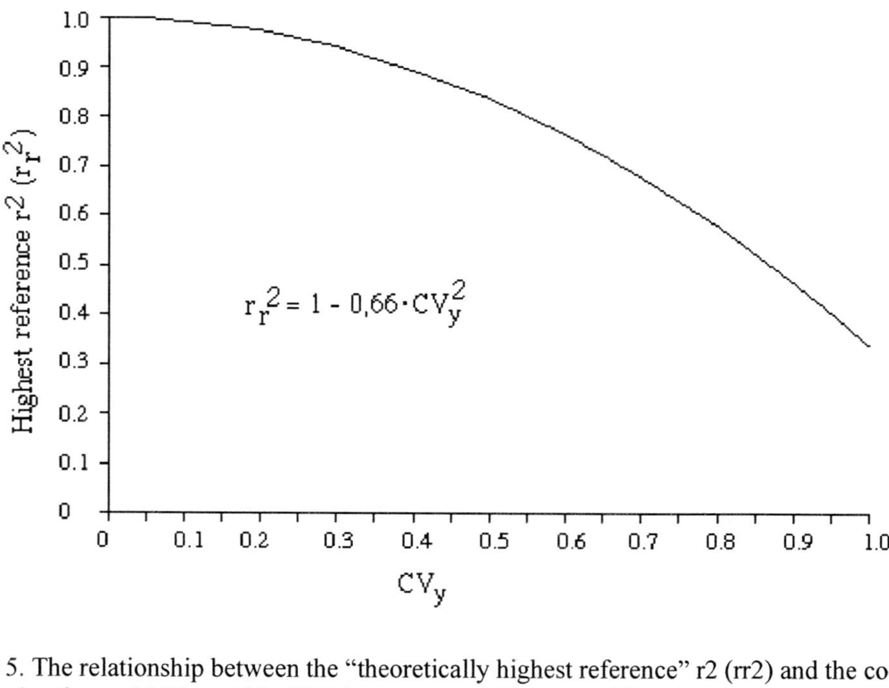

Figure 5. The relationship between the "theoretically highest reference" r2 (rr2) and the coefficient of variation for variability within (CVy) ecosystems. Figure modified from Håkanson (1999).

CV AND SAMPLING PERIOD

The data discussed in this case-study are measurements from the years 1995 to 1999 from the River Danube (see Håkanson et al., 2003 for further information). The water samples were taken at a depth of 5-10 cm. The sample collection frequency varied for different months, higher during the growing season and lower during winter. Up to 20 samples were taken per month, but during some winter months a few or no samples were taken. The water samples have been analysed for different standard algal groups: green algae (incl. *Chlorophyceae, Euglenophyceae* and *Conjugatophyceae*), chromophyte algae (incl., *Bacillariophyceae, Chrysophyceae, Dinophyceae* and *Xanthophyceae*), *Chryptophyceae*, and blue-green algae (cyanobacteria, *Cyanophyceae*). In the River Danube diatoms are the dominant algae among the chromophytes.

Fig. 6A exemplifies all actual (n = 459) data for chlorophyll and fig. 6B gives the same data represented as monthly values with uncertainty bands (the box-and-whisker plots give median values, M50, quartile values, M25 and M75, percentiles, M10 and M90, and outliers). An important question now concerns finding characteristic (mean or median) CV-values based on daily, weekly, monthly and annual samples for the variables in this case-study, i.e., to find the most appropriate CV-values regulating predictive power (r_r^2; eq. 2) or the number of samples according to the sampling formula (eq. 1).

Table 1 shows all chlorophyll data in July for the period 1995 to 1999. This table is included here to stress some important concepts. There are 76 individual samples (given under the column called actual data). In the following, we will calculate weekly, monthly and annual mean values. In table 1, the column marked "MV, n = 6" shows mean values of the actual data for n = 6. One can note that the requested CV for n = 6 is 0.55, which is lower than the CV for the actual data, CV = 0.64. One should also note that the CV of 0.55 is much higher than the CV that can be calculated if the actual data had been randomly distributed in the given time series. Then the CV-value is 0.26. It is a standard statistical practise in many contexts to set the CV for a mean value according to eq. 3, but then one must also, as exemplified in table 1, presuppose that the mean value comes from a random sample from a normal frequency distribution, and the point here is that there are often time trends in variables of this kind so that one cannot use this approach uncritically:

$$CV_{MV} \approx CV_{ind}/\sqrt{n} \quad (3)$$

Where CV_{MV} is the CV for the mean and CV_{ind} is the CV for the individual data; n is the number of data used to determine the mean value.

Table 2 gives a statistical compilation of data for the five algal variables from the Regensburg site. The CV-values from this table are shown in fig. 7. From this figure, one can note that the temporal variations in monthly CVs based on data from several years show similar patterns for the five variables with highest CV during September and October and lowest values during April and May.

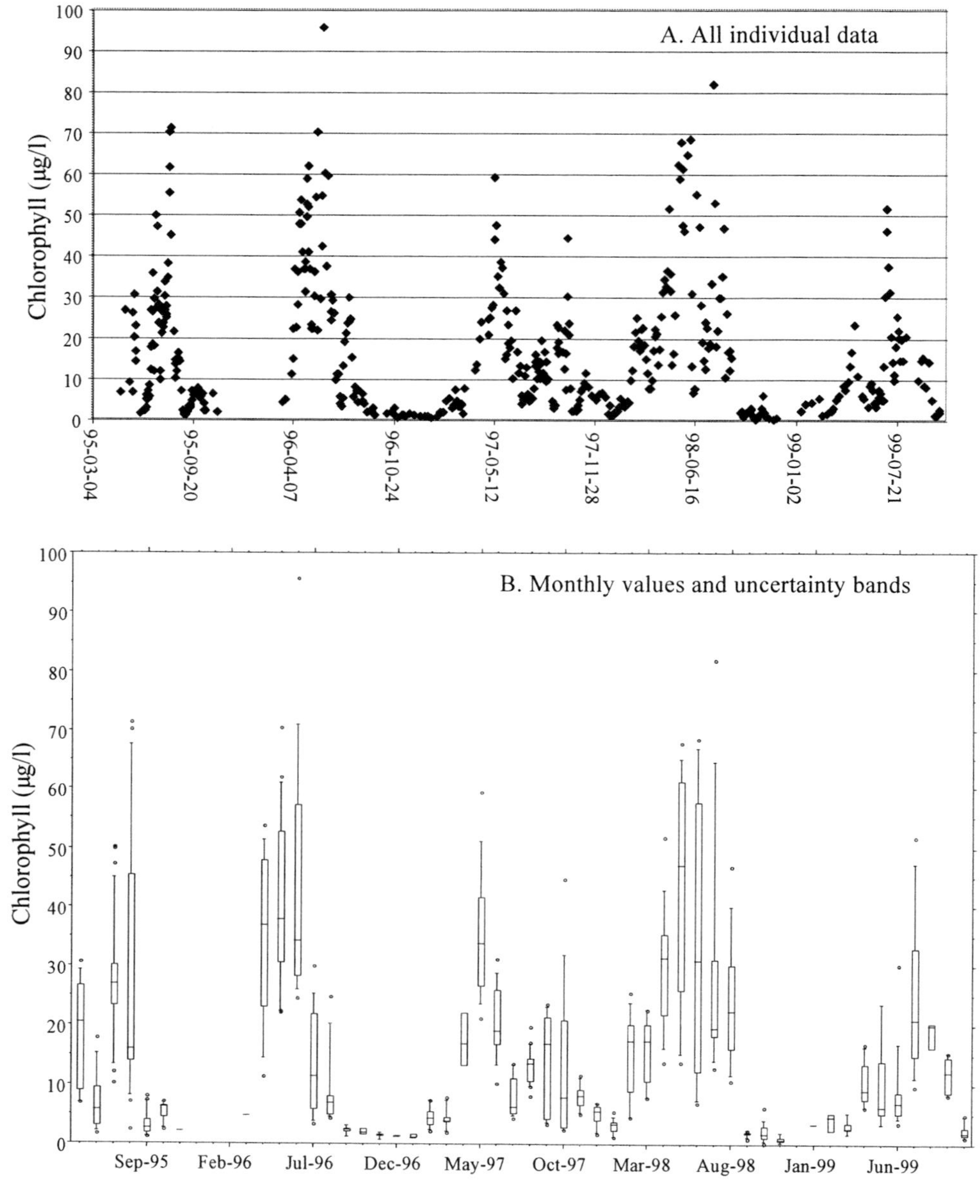

Figure 6. A. All actual data (n = 459) from the sampling site in River Danube (Regensburg station) for chlorophyll concentrations. B. The same data but calculated as mean monthly values and the corresponding uncertainty bands (medians, M50, quartiles, M25 and M75, percentiles, M10 ands M90 and outliers). In this work, we seek to quantify such uncertainties expressed as CV-values for five algal variables (including chlorophyll given here). From Håkanson et al. (2003).

Table 1. All actual data on chlorophyll-a concentrations (µg/l) in July from 1995 to 1999 from the Regensburg station in the River Danube, mean values (MV) from 6 samples, all actual data redistributed randomly, mean values of the redistributed data (n = 6) and mean annual values from this month. The statistics give n = number of data, MV = mean values, SD = standard deviations, CV = coefficient of variations and CV/√6 (data from Håkanson et al. (2003)

Year	Actual	MV	Random	MV, rand	MV	Year	Actual	MV	Random	MV, rand	MV
	data	n = 6	order	n = 6	year		data	n = 6	Order	n = 6	Year
1995	26.7		18.3			1997	11.6		13.6		
	35.8		22.8				13.6	20.7	5.0	9.5	
	26.7		53.2				6.2		6.2		
	18.8		25.0				4.2		5.4		
	12.1		18.3				5.0		27.8		
	18.3	23.0	5.7	23.9			10.9		12.9		
	29.6		18.3				13.2		24.1		
	50.1		5.7				6.5	7.7	14.4	15.1	
	49.9		30.0				5.0		3.4		
	47.3		14.8				5.1		82.0		
	31.5		17.9				6.0		12.1		
	27.7	39.3	23.7	18.4			5.7		19.4		7.7
	28.4		11.4			1998	19.5		6.0		
	23.7		27.7				14.7	9.3	11.6	22.4	
	10.2		13.4				12.9		11.3		
	12.2		33.4				24.0		28.4		
	21.4		21.4				22.8		26.8		
	26.4	20.4	46.2	25.6			17.9		37.7		
	26.7		4.0				18.9		22.1		
	22.8		29.6				33.4	21.7	20.8	24.5	
	23.8		19.5				82.0		4.2		
	30.3		22.8				53.2		51.7		
	33.9		9.8				18.3		31.5		28.9
	26.8	27.4	26.7	18.7		1999	46.2		30.3		
	26.0		26.4				51.7		24.0		
	25.0		9.9				37.7	48.2	25.6	27.9	
	27.8		26.0		27.4		31.4		26.7		
1996	9.9		35.8				20.8		11.3		
	11.3		33.9				14.4		14.7		
	11.3	18.5	23.8	26.0			9.8		21.3		
	5.7		18.8				11.4		5.1		
	4.0		20.3				18.3	17.7	18.9	16.3	
	3.4		6.5				25.6		49.9		
	5.4		50.1				20.3		31.4		
	13.4		26.7				22.1		47.3		
	19.4	8.5	23.9	24.4			14.8		13.2		25.0
	21.3		5.0			**n**	**76**	**12**	**76**	**12**	**5**
	23.9		10.2			**MV**	**21.81**	**21.87**	**21.81**	**21.05**	**20.61**
	30.0		12.2			**SD**	**14.03**	**12.06**	**14.03**	**5.46**	**9.25**
	24.1		10.9		14.1	**CV**	**0.64**	**0.55**	**0.64**	**0.26**	**0.45**
						CV/√6	**0.26**				

Table 2. Monthly data from 1995 to 1999 from the Regensburg station (the Danube). The statistics give n = number of data, MV = mean values, SD = standard deviations, CV = coefficients of variations

A. Chl	Jan	Feb	Mar	April	May	June	July	Aug	Sep	Oct	Nov	Dec
N	17	18	25	34	54	58	76	58	46	46	18	9
MV	2.62	9.58	9.06	26.69	33.57	21.88	21.81	19.65	6.67	6.63	3.84	3.72
SD	1.33	7.57	7.10	13.99	17.64	20.16	14.03	15.72	6.74	9.79	3.67	2.39
CV	**0.51**	**0.79**	**0.78**	**0.52**	**0.53**	**0.92**	**0.64**	**0.80**	**1.01**	**1.48**	**0.95**	**0.64**
B. Greens												
N	17	18	25	32	44	45	69	50	41	42	18	8
MV	0.42	1.36	1.40	3.11	5.21	6.22	4.68	4.59	1.94	1.38	0.56	0.55
SD	0.19	1.03	0.98	1.41	3.01	5.40	2.82	3.29	2.17	2.21	0.45	0.34
CV	**0.45**	**0.76**	**0.70**	**0.45**	**0.58**	**0.87**	**0.60**	**0.72**	**1.12**	**1.61**	**0.81**	**0.62**
C. Diatoms												
N	17	18	25	32	44	45	69	50	42	47	18	8
MV	1.87	7.30	6.84	19.52	27.26	17.14	15.63	11.09	4.39	4.21	2.69	2.77
SD	1.04	5.81	5.30	10.20	13.30	15.43	12.51	9.72	4.23	5.59	2.73	1.64
CV	**0.56**	**0.80**	**0.77**	**0.52**	**0.49**	**0.90**	**0.80**	**0.88**	**0.96**	**1.33**	**1.01**	**0.59**
D. Cryptophyta												
N	17	18	25	32	44	45	59	44	40	42	18	8
MV	0.26	0.71	0.67	1.01	2.55	1.79	1.19	1.08	0.65	0.49	0.47	0.56
SD	0.13	0.66	0.61	0.92	1.52	1.20	0.88	0.71	0.53	0.45	0.45	0.35
CV	**0.48**	**0.93**	**0.92**	**0.91**	**0.60**	**0.67**	**0.74**	**0.65**	**0.82**	**0.90**	**0.96**	**0.62**
E. Blue-greens												
N	17	18	25	31	44	45	50	43	32	40	17	8
MV	0.06	0.20	0.25	0.83	1.26	0.78	0.32	0.41	0.36	0.29	0.12	0.14
SD	0.04	0.19	0.21	0.61	0.91	0.71	0.26	0.35	0.55	0.55	0.11	0.09
CV	**0.71**	**0.93**	**0.86**	**0.73**	**0.72**	**0.91**	**0.81**	**0.86**	**1.52**	**1.90**	**0.92**	**0.66**

Fig. 8 gives a compilation of CV-values calculated for mean values from weekly, two types of monthly values (individual monthly data and monthly data based on values from several years) and yearly values, as well as the CV-value for all 459 individual data. The latter value is 0.96 and in fig. 8 we compare this value to 19 similar values for chlorophyll for 19 sites in UK rivers. The UK data-base (see Foster et al., 1996; 1997; 1998) provides a range of data for 79 monitoring points in the catchment of the Warwickshire River where the mean annual flow is about 15 m^3/s. Around 900,000 people live in the catchment. These sampling stations included effluent discharges from 18 sewage treatment works; 57 river sites with catchments ranging in area from 50 km^2 to 2200 km^2; and 4 lakes/reservoirs. Sampling was undertaken between December 1994 and September 1996 and sampling frequencies ranged from weekly to monthly.

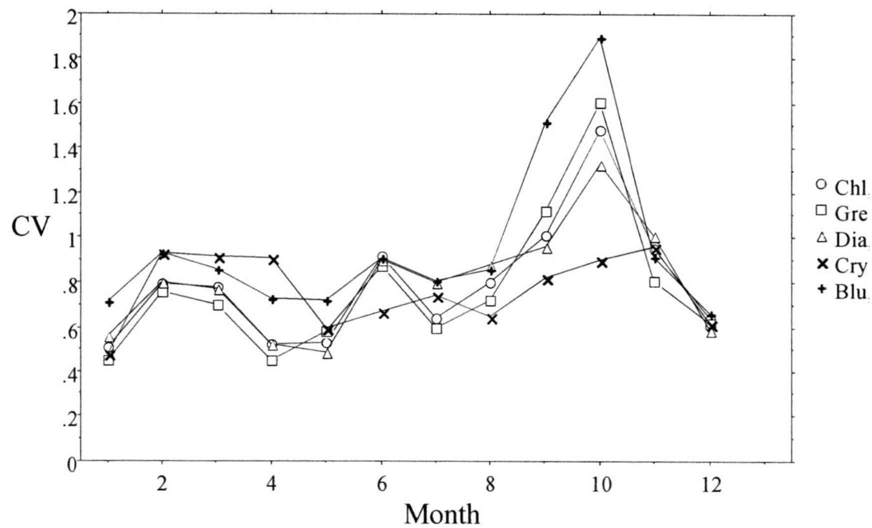

Figure 7. Monthly CV-values for the five different algal variables based on data from 1995-1999 from the River Danube (Regensburg station; from Håkanson et al., 2003).

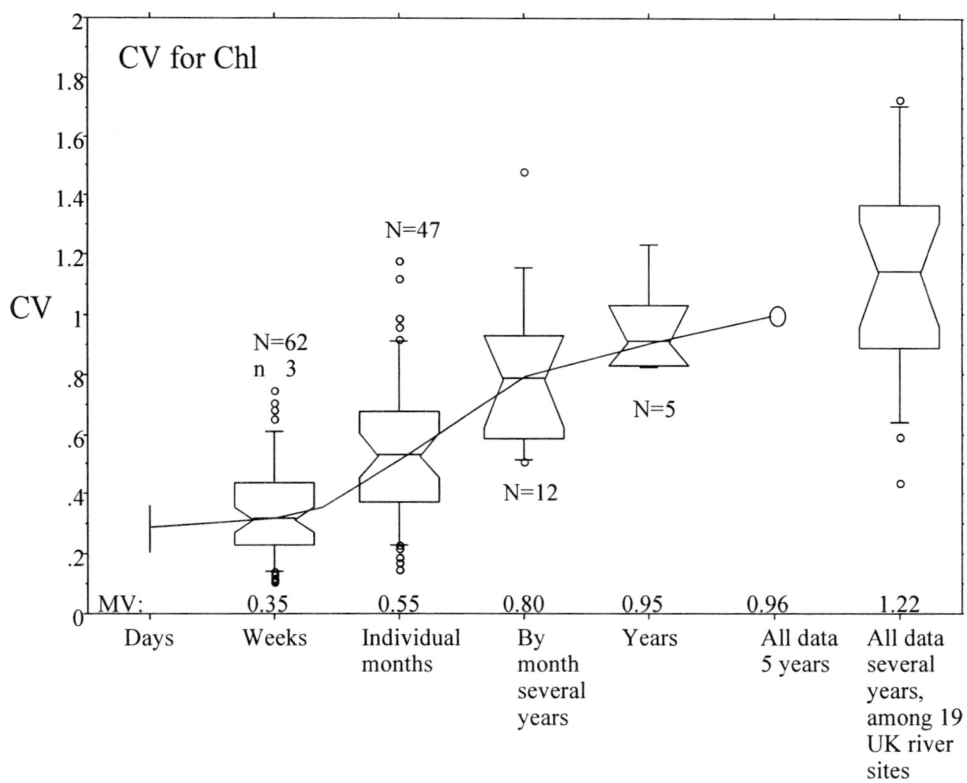

Figure 8. Compilation of CV-values based on mean values from different time intervals, daily, weekly, monthly (from individual months and monthly data from several years), yearly and all data from 1995-1999 for chlorophyll from the River Danube (Regensburg station). The box-and-whisker plot to the right gives for comparative purposes CV-values based on all data from several years from 19 UK rivers. The results for the UK rivers sites should be compared to the CV-value of 0.96 calculated using all data from the Regensburg station (from Håkanson et al., 2003).

From fig. 8, one can note:

- The mean CV for chlorophyll based on all data from the site in River Danube is close to the median value from 19 rivers sites in the UK.
- The daily, weekly, monthly (based on individual months), monthly (based on data from several years) and yearly CVs increase in steps from 0.3, 0.35, 0.55, 0.80 to 0.95, respectively for chlorophyll.
- Analogous results are obtained also for green algae, diatoms, cryptophytes and blue-greens, and basically for all water variables.
- The result for the daily CV-value emanate from a simple extrapolation based on the median values in the box-and-whisker plots, so this value may be regarded as a first indication of a likely order-of-magnitude value for a characteristic CV for chlorophyll at this time scale.

Corrections and Normalizations

One can also ask if it is possible to reduce CV by various types of corrections. We have already discussed the evident approach to calculate CV not from individual samples but from mean values (eq. 3) and in this section we will see if there exists a relationship between the CVs and the mean values (MV); then we will see if there is any relationship between CV and flow velocity. We will also test if there are any relationships between the mean/median (or characteristic) CV and various maximum values for CV.

CV Versus Mean Values

A low mean value may results in a high CV-value due to the definition of CV as SD/MV (if SD is constant). It is therefore interesting to see if there exists a "border concentration" below which the CV-value starts to increase. A scatter plot between CV and the corresponding mean values will show if such a concentration exists. As shown in fig. 9, CV does not increase with decreased MV and therefore no significant border concentration can be identified neither for chlorophyll, diatoms nor for any of these algal variables in River Danube and probably not for most water variables in lakes and rivers.

Characteristic CV and Dependence of Flow Velocity

Since these data emanate from a river, the flow velocity could be expected to be of specific importance (Swale, 1964; Bodemer, 1998). Fig. 10 shows the relationship between monthly CV for chlorophyll and the corresponding mean flow velocities for this river site. The scatter plot and the regression show that the r^2-value is very low (0.026) and not significant ($p = 0.32$). This suggests that flow velocity cannot be used to reduce the CV for chlorophyll significantly. This is probably true also in other rivers and most water variables.

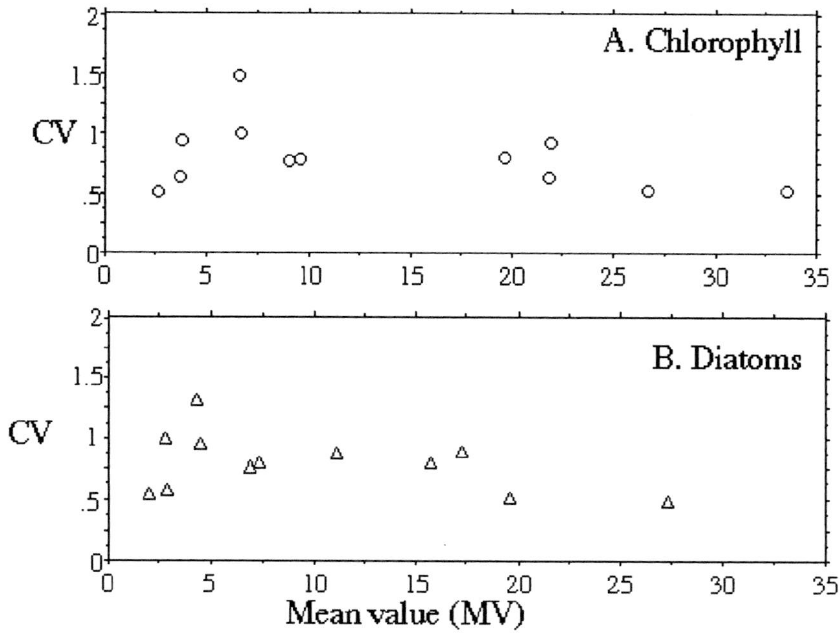

Figure 9. The coefficients of variation (CV) based on data from individual months plotted against the corresponding mean values (MV) for chlorophyll and diatoms from the River Danube (Regensburg station; data from 1995-1999). Data from Håkanson et al. (2003).

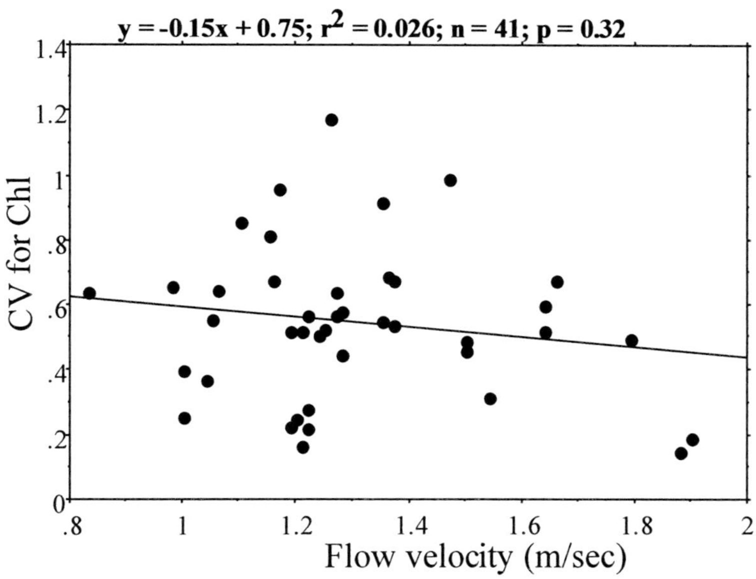

Figure 10. The relationship between the CV for chlorophyll calculated for individual months and the corresponding mean flow velocity and statistics (regression line, r2; n and p) using data from the River Danube (Regensburg station; data from 1995-1999 from Håkanson et al., 2003).

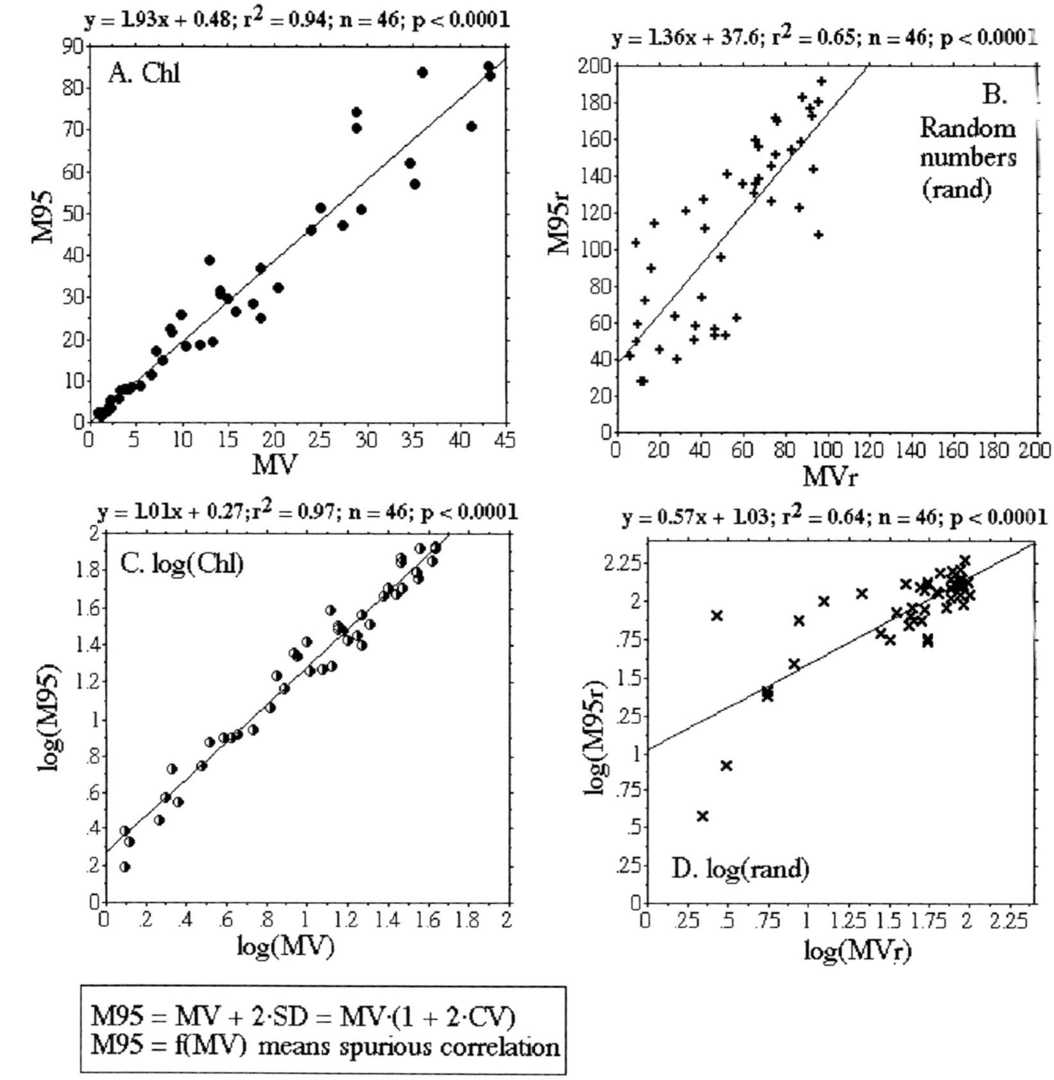

Figure 11. Four regressions between M95 (the 95-percent value ≈ MV + 2·SD; on the y-axis) and mean monthly values of chlorophyll for individual months (MV on the x-axis) using data from the River Danube (Regensburg station; data from 1995-1999). A. Gives the regression using actual values for chlorophyll. B. Gives a corresponding regression using data generated by a random number generator. This is highly significant ($r^2 = 0.65$; $p < 0.0001$) but entirely spurious. C. Gives the regression using logarithmic values for chlorophyll. D. Gives the corresponding regression using logarithmic values generated by a random number generator. This is also a spurious regression. From Håkanson et al. (2003).

Mean Versus Maximum Values

For many purposes, the mean values are of interest, but it is important also to forecast extreme concentrations, and extreme algal bloom events. One can use the data-base from river Danube to illustrate this question (see fig. 11). The results demonstrate a strong logical correlation between the monthly mean values and the different peak values. This is exemplified for M95-values (i.e., 95th percentile values) and mean monthly values (MV) for chlorophyll in fig. 11. The r^2-value is 0.94 for the actual data (n = 46; see fig. 11A) and 0.97 for the logarithmic data (fig. 11D). This is in good general agreement with results presented for lakes by Vollenweider and Kerekes (1980). Partly, this can be explained by the fact that all the distributions are approximately normal after logarithmic transformations. This high correlation is not, however, self-evident since several factors, like self-shading and grazing, affect the variation of phytoplankton concentrations. Partly, the excellent correlation shown in fig. 11 can also be attributed to "spurious correlation" (see Håkanson and Boulion, 2002).

Spurious Correlations

For a normal (Gaussian) frequency distribution, we have:

$$M95 \approx (MV + 2 \cdot SD) \quad (4)$$

This means that the M95-values by definitions include information on the mean value (MV). Using randomly generated data, one can illustrate a very important concept in statistical analysis, spurious correlations. First, 46 randomly data, called MVr, have been generated, then 46 random data called SDr. From these two data-sets, one can calculate M95r = MVr + 2·SDr. The correlation between M95r and MVr is shown in fig. 11B for the actual values; the r^2-value is 0.65, which is highly significant, but totally spurious. The correlation for the logarithmic values are given in fig. 11D (r^2 = 0.64). This is a mathematical consequence of regressing a variable against an expression including the variable. This is well-known and basic in statistics and mathematics, but nevertheless sometimes a practice in aquatic sciences.

Model Testing

Calibration and Validation

Before it is meaningful to critically test a model, it should be calibrated. Calibration means that a given model set-up is tested against empirical data so that the fit between modelled values and empirical data becomes as good as possible. One example of this is illustrated in fig. 12 using a simple dynamic mass-balance model (see fig. 13) for a Swedish lake, Lake S. Bullaren (see Håkanson, 1999) to test which value for the internal loading rate (R_{int}) for phosphorus which gives the best correspondence between modelled values and empirical data for total phosphorus (TP) concentrations in water and surface sediments. If the internal loading (both resuspension and diffusion in this simplified model) is set too high, too much phosphorus will be found in the water and too little will remain in the sediments, and vice versa. It is evident that a R_{int}-value of 0.2 per year (i.e, 20% of the phosphorus deposited in the sediments will leave the sediments each year) provides the best fit in this example.

One should also note that R_{int} is the main target in calibrating this model. The settling velocity regulating sedimentation is set to 72 m/yr, the inflow is calculated from empirical data on the TP-concentration in the tributaries (mean value = 60 μg/l), the outflow is regulated by the theoretical lake water retention time (T; calculated on an annual basis in this model), and the burial by setting the active sediment layer to 4 cm and the mean deposition to 4 mm/yr (see Håkanson and Peters, 1995). This gives a mean age of the sediments of 10 year, which regulates the burial rate (see fig. 13).

All model variables (such as rates and distribution coefficients) could and should be tested like this, and there are generally several combinations of values for the model variables that can give good predictions when calibrated against a set of empirical data in one lake or river. All such combinations cannot be correct if one seeks the model constants which could be used as general default values in model simulations for many lakes.

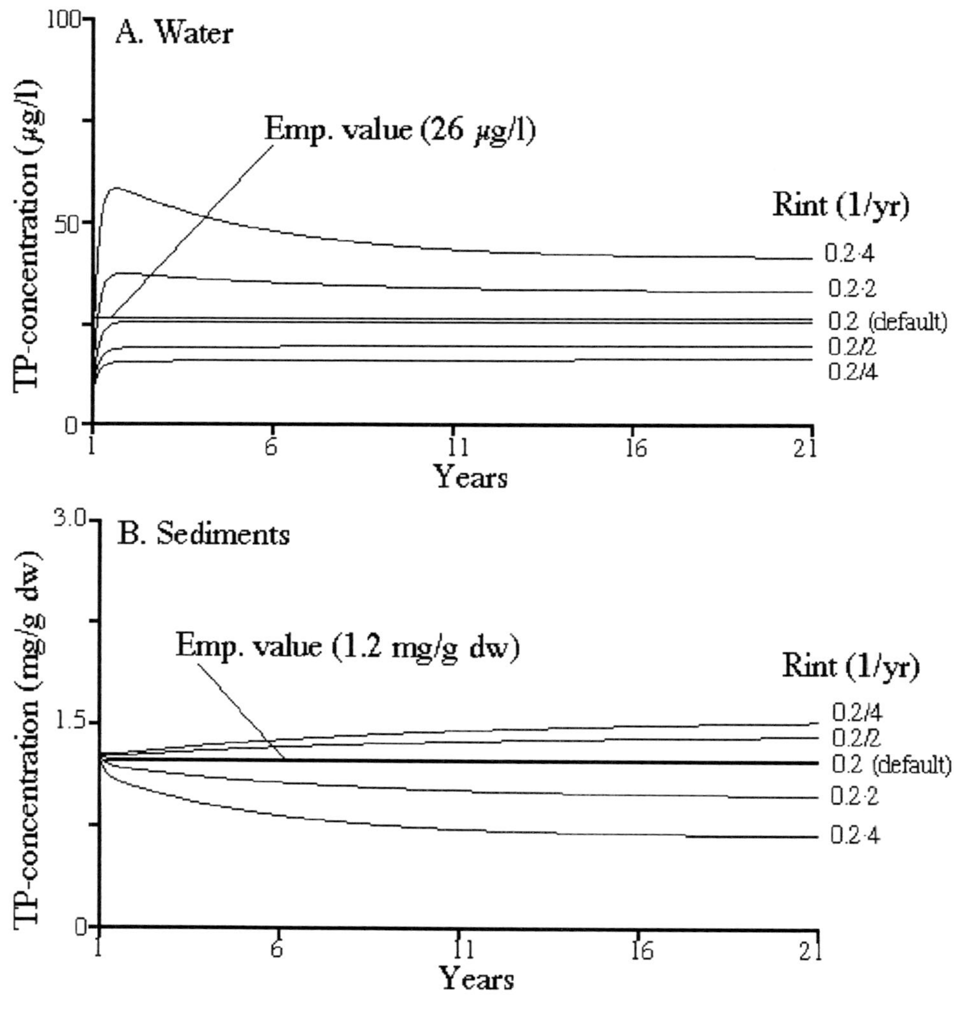

Figure 12. Illustration of a calibration. Different values for the rate regulating internal loading of phosphorus in the mass-balance model (Rint) in Lake S. Bullaren, Sweden, are tested to see which value can best maintain the empirical concentrations of phosphorus in lake water and sediments. The empirical target value for lake water is about 26 μg/l and for lake sediments, the empirical target is 1.2 mg/g dw. These values are best predicting the Rint is set to 0.2 (1/yr). It the internal loading is higher, water concentrations will be too high and sediment concentrations too low, and vice versa.

This means that calibrations often involve many iterations. The idea with the calibration procedure is that for each round of iterations the uncertainty in the values for the model variables should be reduced. When the model is properly calibrated, it should be validated, i.e., tested against independent data. It is evident that it is preferable if the calibrations and the validations include as reliable empirical data from as many systems as possible covering as wide a range as possible in model variable characteristics.

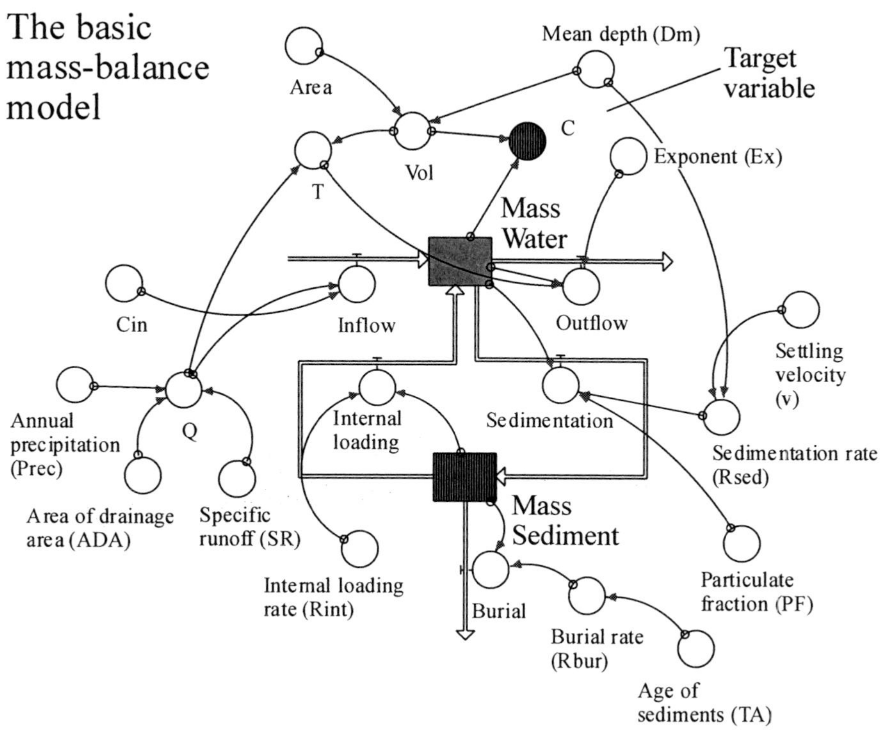

MS(t) = MS(t - dt) + (FWS - FSW - Fbur)·dt

FWS = MW·PF·Rsed
FSW = MS·Rint
Fbur = MS·Rbur

MW(t) = MW(t - dt) + (Fin + FSW - Fout - FWS)·dt

Fin = Q·Cin
FSW = MS·Rint
Fout = MW/T^Ex
FWS = MW·PF·Rsed

ADA = 199 km2
Area = 8.28 km2
C = 1000·MW/Vol
Cemp = 26 μg/l
Cin = 60·0.001
Dm = 10.1 m
Ex = 1 (default value)
PF = 0.58(default value)
Prec = 850 mm/yr
Q = SR·60·60·24·365·ADA·(Prec/650) m3/yr
Rbur = 1/TA 1/yr
Rint = 0.2 1/yr
Rsed = v/Dm 1/yr
SR = 0.01 m3/km2·sec
T = Vol/Q yr
TA = 10
v = 72 m/yr (default value)
Vol = Dm·Area m3

Figure 13. An extended version of the basic mass-balance model and a simplified version of the comprehensive model. Equations, target variables, lake-specific variables (for Lake S. Bullaren, Sweden), and model variables

To illustrate a validation and some inherent problems with simple models in lake management (such as commonly used models of the Vollenweider (1968) and OECD-type; OECD, 1982, which do not account for internal loading and burial), fig. 14 gives a comparison between model-predicted lake concentrations of total phosphorus, C_{TP} (using the OECD-model), and measured C_{TP} for 18 lakes of varying limnological character in the domain of the OECD-model (see table 3). The OECD-model is illustrated in fig. 15 and given by:

$$C_{TP} = C = 1.55 \cdot (C_{in}/(1+T))^{0.82} \quad (5)$$

Where $C_{in} = C_{TPin}$ is the mean annual TP-concentration (µg/l) in the tributary and T is the theoretical lake water retention time (years).

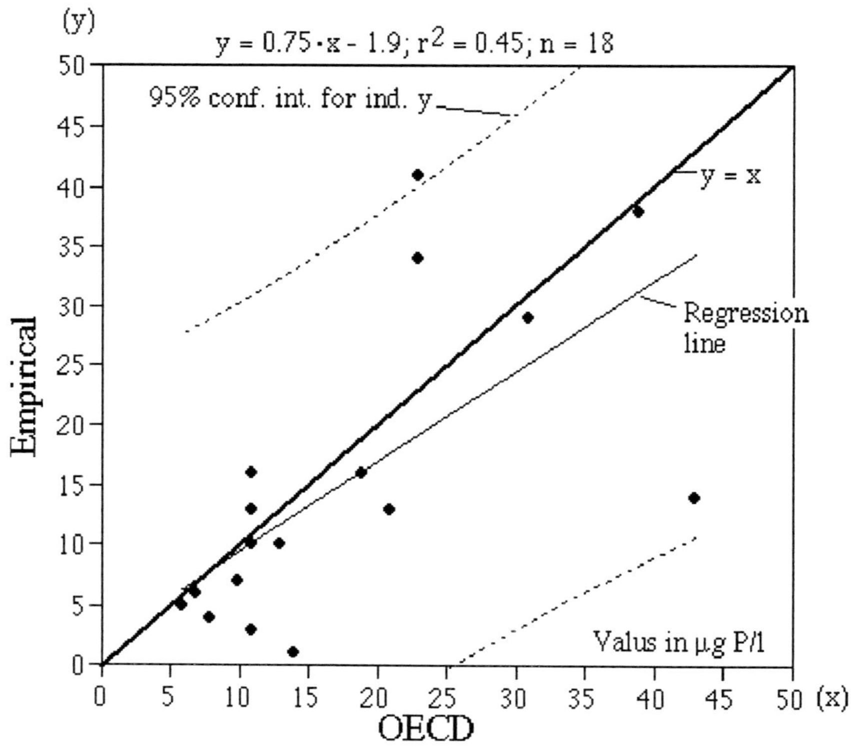

Figure 14. Illustration of a validation of the OECD-model against an independent set of data from 18 lakes

The lakes in this test were not included in the development of the OECD-model. The validation gives an r^2 of 0.45 for actual values of C_{TP} and 0.4 for logarithmic values. Fig. 14 also shows the 95% confidence intervals for the individual y-values for this regression. This validation indicates that the uncertainty associated with the OECD-model is so large that the model is unsuitable for predicting C_{TP} in many lakes. Models of this type can give good predictions for many lakes, even for most lakes (within the domain of the model), but it is important to note that it is often not possible to predict for which lakes the models do and do not work! Basically, these models should not be used for lakes dominated by internal loading, but it may be difficult to develop operational criteria for this, and it is also difficult to develop criteria for the level of statistical uncertainty that should be applied for the driving variables, especially the TP-concentration in the tributary water. How many samples are actually needed to determine the mean TP-concentration for the tributaries to a given lake? It should also be noted that:

1. In spite of the fact that the model variables (in this case the theoretical lake water retention time and the TP-concentration in the tributary) fall in the model domain, some of the lakes used in this validation may have characteristics outside the range of the OECD-model. For example, the lake type may not be appropriate.
2. Many of the lakes used for this validation (see table 3) probably have a significant internal loading of phosphorus (from resuspension and diffusion). There is a close relationship between internal loading, bottom dynamic conditions and the dynamic ratio (DR = √Area/Dm; Area in km^2 and mean depth in m; see Håkanson and Jansson, 1983). If DR is larger than about 2, more than 50% of the lake is likely to be dominated by areas of erosion and transportation (ET-areas). In such lakes, resuspension from wind/wave action is a most important factor for lake TP-concentrations. From table 3, one can note that there are several lakes with DR-values larger than 2.
3. The high r^2-value (0.86, see fig. 15) obtained in the derivation of the OECD-model can, probably, partly be attributed to the fact that the data used in that work emanate from a scientific project where great efforts were made to ensure the quality of the data. Most data used in practical water management are probably less reliable. This has strong bearings on the predictive success. Many of the values in table 3, which emanate from an international data base, are, in all likelihood, of rather poor quality. This is an often encountered problem, especially for many data-bases available on the Internet, and this problem should not be disregarded.

Sensitivity Tests

Sensitivity analysis involves the study of how an alteration of one model variable influences a given prediction, while everything else is kept constant. This analysis plays an important role in ecosystem modelling (see Hinton, 1993; Hamby, 1995; IAEA, 1998). This section gives an example of how a sensitivity analysis can be performed. The calibration in fig. 12 can also be seen as a sensitivity analysis. The R_{int}-value has been varied and all else kept constant in a simulation to see how variations in the given model variable influence a given target variable. However, sensitivity analysis is a wider concept and it usually involves, at least, two further steps.

Fig. 16 gives a sensitivity analysis where the default value of the settling velocity, v (72 m/yr), from the model in fig. 13 has been changed 100 times while all else in the model have been kept constant. It has also been assumed that there exist a typical, characteristic mean value for v (here 72 m/yr) and a given uncertainty for this value given by a standard deviation, which has been set to 25% of the mean. That is, the CV-value has been set to 0.25 for v. From a frequency distribution with a mean value of 72 and a standard deviation of 18,

100 data have been selected at random and used in the model to produce the 100 curves for the target variable, the lake TP-concentration. It is evident from fig. 16 that the predictions of the y-variable are sensitive to the value selected for the settling velocity (v). The uncertainty in y is also shown by the box-and-whisker plot in fig. 16 showing the median value (M50), quartiles (M25 and M75), percentiles (M10 and M90) and outliers. One can note that when steady-state has been reached, the CV-value for y is 0.16 related to the uncertainty in the given x-variable (here v).

Table 3. Lake data from an international data register. T = theoretical lake water retention time; Q = mean annual water discharge; D_{max} = maximum depth.; D_m = mean depth; DR = dynamic ratio; C_{TPin} = mean annual total-P concentration in tributaries; C_{TP} = mean annual total-P concentration in lake water; OECD = mean annual C_{TP} calculated with the OECD-model. Data from Meeuwig and Peters (1995).

Lake (nr)	Area (km^2)	Vol ($m^3 \cdot 10^6$)	T (yr)	Q (10^6 m^3/yr)	D_m (m)	D_{max} (m)	DR	C_{TPin} (µg/l)	C_{TP} (µg/l)	OECD (µg/l)
236	80.3	366	0.25	1464	4.6	6.4	1.97	77	38	39
241	1140	13600	2.2	6182	11.9	61	2.83	95	29	31
242	478	2900	3.3	879	6.1	22	3.60	74	41	23
243	1856	74000	55.9	1324	39.9	128	1.08	49	6	7
244	5648	153000	9	17000	27.1	106	2.77	52	10	13
246	867	8500	1.9	4474	9.8	60	3.00	25	10	11
247	1100	17800	2.7	6593	16.2	98	2.05	30	13	11
248	13.5	206	3.5	59	15.3	87	0.24	68	13	21
250	117.5	887	0.15	5913	7.5	36	1.44	36	34	23
261	1053	11900	0.25	47600	11.3	63.1	2.87	9	6	7
264	42.3	1640.1	13.4	122	38.8	84.1	0.17	49	16	11
271	3.8	55	0.47	117	14.5	30.2	0.13	26	1	14
272	7.4	118	0.47	251	15.9	29.3	0.17	12	4	8
277	11.5	1160	6.1	190	100.9	163	0.03	37	3	11
295	4.25	189.1	0.68	278	44.5	142.5	0.05	18	7	10
296	5.96	55.5	0.31	179	9.3	16.1	0.26	89	14	43
301	616	60000	10.6	5660	97.4	164	0.25	20	5	6
306	65.1	3300	1.1	3000	50.7	136	0.16	43	16	19
Ranges for		**Min.**	**0.1**					**1**		
OECD-model variables:		**Max.**	**100**					**150**		

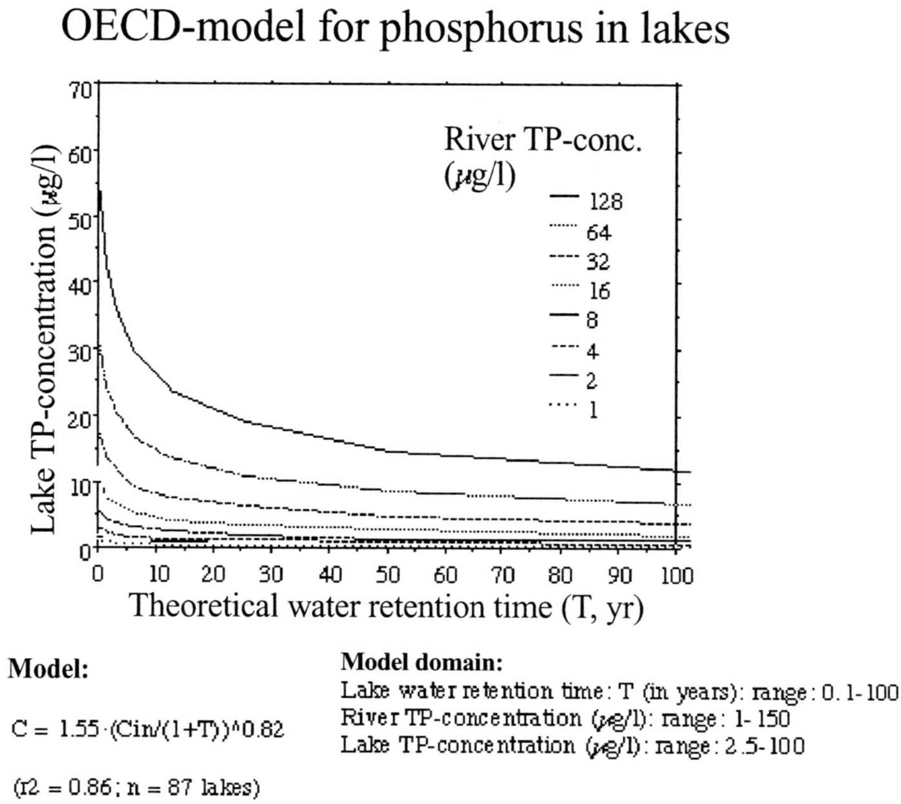

Figure 15. Illustration of the well-known OECD (1982) model.

The next step in a sensitivity analysis is often to repeat this type of calculation for all interesting model variables to try to produce a ranking of the factors influencing the target variable. The basic idea is to identify the most sensitive part of the model, i.e., the part that is most decisive for the model prediction. An example of such a comprehensive sensitivity analysis is shown in fig. 17. In this case, the following major fluxes in the model were selected for the test: Tributary TP-concentration (C_{in}; fig. 17A), sedimentation (B), outflow (C), internal loading (D), and, finally, burial (E). These analyses have been done for the target y-variable, the lake TP-concentration. In these sensitivity analyses, all model variables were altered by the same factor (2). So, this is a sensitivity test for fluxes using uniform uncertainties. From fig. 17, one can note that from these presuppositions, the TP predictions are most sensitive to the choice of the values for inflow (C_{in}), sedimentation (v) and internal loading (R_{int}). This is very logical and applies for many substances in many types of aquatic systems because these three model variables regulate two primary fluxes, inflow (C_{in}) and sedimentation (v). All other internal fluxes in lakes depend on these fluxes.

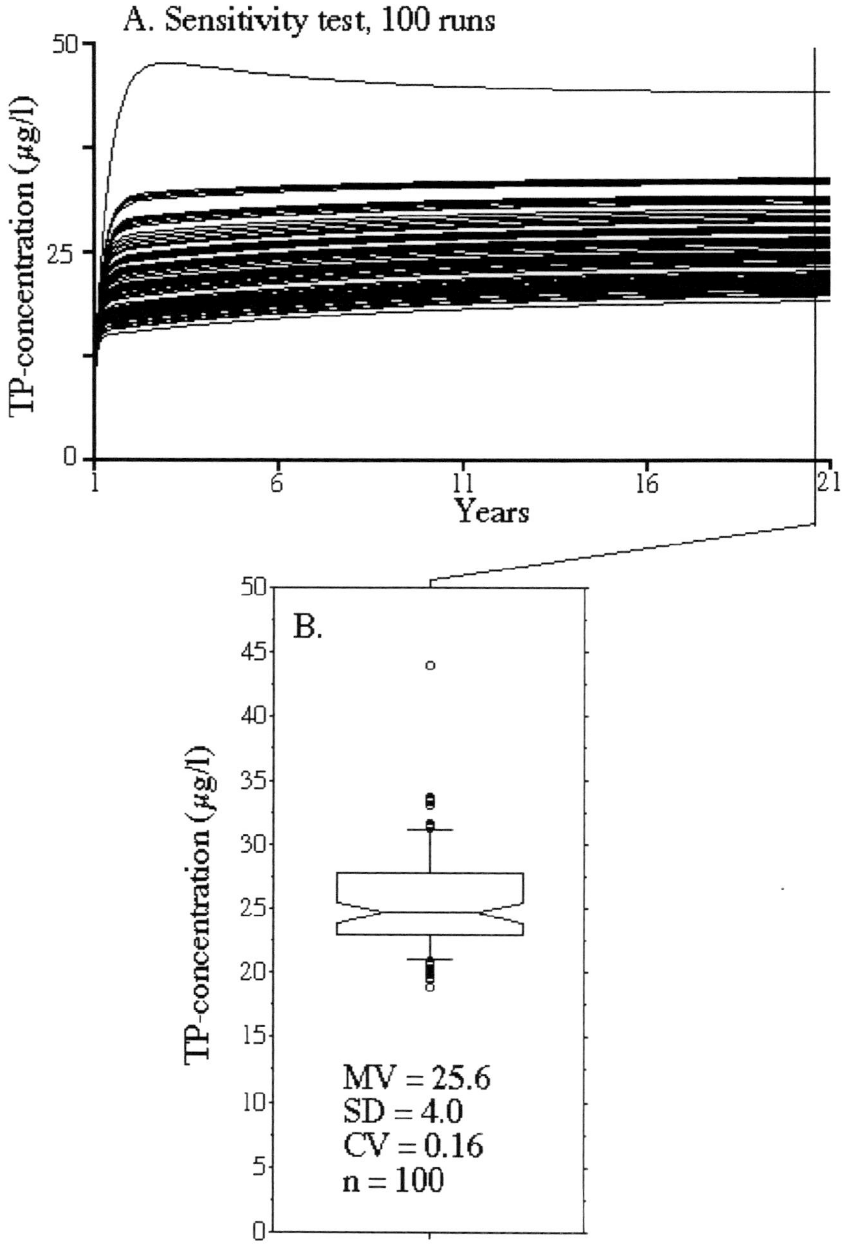

Figure 16. Sensitivity analysis using the mass-balance model for Lake S. Bullaren, Sweden. The target variable is the lake TP-concentration and the settling velocity (v) has been varied (mean value 72, CV = 0.25). 100 runs. The figure also gives the box-and-whisker plot for the data from year 20.

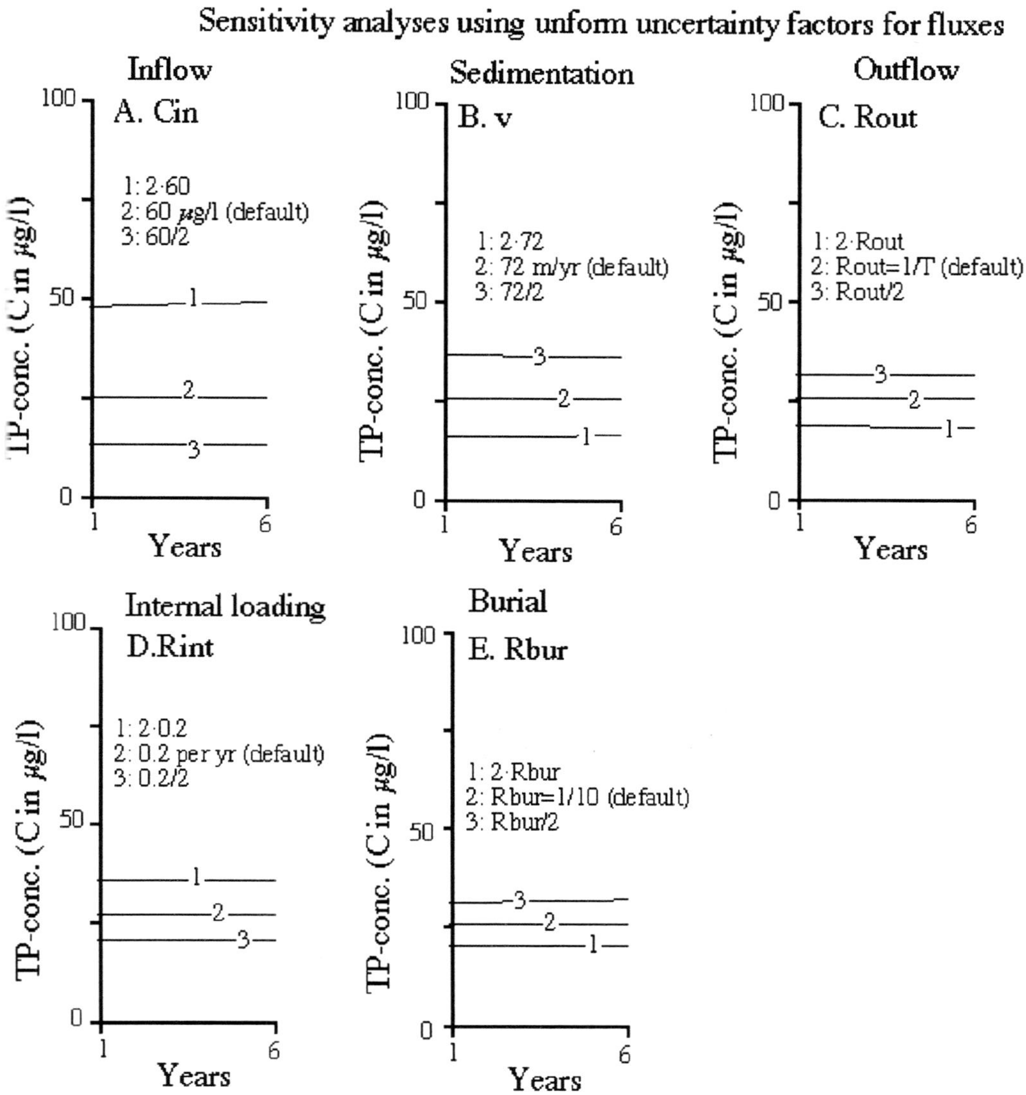

Figure 17. Results for sensitivity analyses using uniform uncertainties (a factor of 2) for all fluxes in the mass-balance model applied for Lake S. Bullaren, Sweden

However, it is evident that it is not realistic to apply the same uniform uncertainty for all model variables, such as a factor of 2 in the given example. There are major differences among model variables in this respect. Morphometric parameters can often be determined very accurately (see Pilesjö et al., 1991), some model variables, like rates and distributions coefficients, can, on the other hand, not be empirically determined at all for real water systems, but have to be estimated from laboratory tests or theoretical derivations. This means that the values used for such model variables are often very uncertain. Table 4 gives a compilation of typical, characteristic CV-values for many types of lake variables used in aquatic modelling. Note that in rivers, one should expect the CV for a given variable to be a factor of 1.5 to 2 higher than in lakes. From table 4, one can note that model variables like rates and distribution coefficients generally can be given CV-values of 0.5. The highest expected CV-values appear for certain sedimentological variables, like TP-concentration in T- and E-sediments. In the following uncertainty tests, we will use the CV-values given in table 4.

Table 4. Compilation of characteristic CV-values for different types of lake variables. Note that the CV-values are likely a factor of 1.5 to 2 higher for river samples compared to lake samples (from Håkanson, 1999)

	CV		CV
Catchment variables		**Lake management variables**	
Catchment area (ADA)	0.01	Secchi depth (Sec)	0.15
Percent outflow areas (OA)	0.10	Chlorophyll-a concentration (Chl)	0.25
Mean soil type or permeabilty factor (SP)	0.25	Cs-concentration in fish	0.22
		Cs-concentration in water	0.30
Lake variables		Cs-concentration in sediments	0.60
Lake area (Area)	0.01		
Mean depth (Dm)	0.01	**Climatological variables**	
Volume (Vol)	0.01	Annual runoff rates	0.10
Maximum depth Dmax)	0.01	Annual precipitation	0.10
Theoretical water retention time (T_w)	0.10	Temperatures	0.20
Water chemical variables		**Model variables**	
pH	0.05	Fall velocities	0.50
conductivity (cond)	0.10	Age of A-sediments	0.50
Ca-concentration (Ca)	0.12	Age of ET-sediments	0.50
Hardness (CaMg)	0.12	Diffusion rates	0.50
K-concentration (K)	0.20	Retention rates	0.50
Colour (col)	0.20	Bioconcentration factors	0.50
Fe-concentration (Fe)	0.25	Feed habit coefficients	0.50
Total-P concentration (TP)	0.35	Distribution (=partion) coefficients	0.50
alkalinity (alk)	0.35		
Sedimentological variables			
Percent ET-areas (ET)	0.05	Mean TP-conc. for E-areas	0.50
Suspended particulate matter conc. (SPM)	0.20	Mean TP-conc. for T-areas	0.75
Mean water content for E-areas	0.30	Mean TP-conc. for A-areas	0.35
Mean water content for T-areas	0.20	Mean metal conc. for E-areas	0.50
Mean water content for A-areas	0.05	Mean metal conc. for T-areas	0.25
Mean bulk density for E-areas	0.10	Mean metal conc. for A-areas	0.20
Mean bulk density for T-areas	0.10		
Mean bulk density for A-areas	0.02		
Mean organic content for E-areas	0.50		
Mean organic content for T-areas	0.50		
Mean organic content for A-areas	0.10		

Uncertainty Tests Using Monte Carlo Techniques

Two main approaches to uncertainty analysis exist, analytical methods (Cox and Baybutt, 1981; Beck and Van Straten, 1983; Worley, 1987) and statistical methods, like Monte Carlo techniques (Tiwari and Hobbie, 1976; Rose et al., 1989). In this section, we will only discuss Monte Carlo simulations. Uncertainty tests using Monte Carlo techniques may be done in several ways, using uniform CV-values, or more realistically, using characteristic CV-values (e.g., from table 4). For predictive models based on several uncertain model variables (rates, etc.), the uncertainty in the prediction of the target variable (y) depends on such uncertainties. The cumulative uncertainty from many uncertain x-variables may be calculated by Monte Carlo simulations.

Monte Carlo simulations is a technique to forecast the entire range of likely observations in a given situation; it can also give confidence limits to describe the likelihood of a given event. Uncertainty analysis (which is a term for this procedure) is the same as conducting sensitivity analysis for all given model variables at the same time. A typical uncertainty analysis is carried out in two steps. First, all the model variables are included with defined uncertainties and the resulting uncertainty for the target variable calculated. Then, the model variables are omitted from the analysis one at the time. The procedure is illustrated in fig. 18.

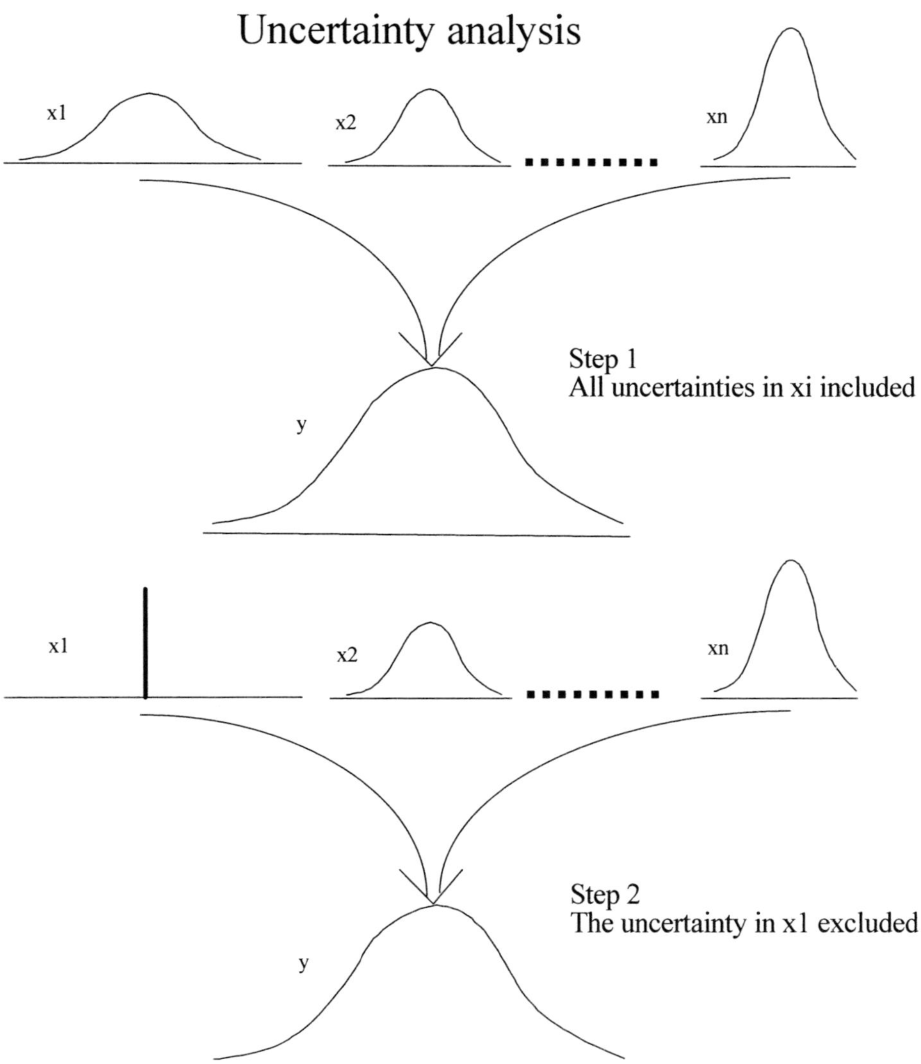

Figure 18. Illustration of the principles of uncertainty analysis using Monte Carlo simulations.

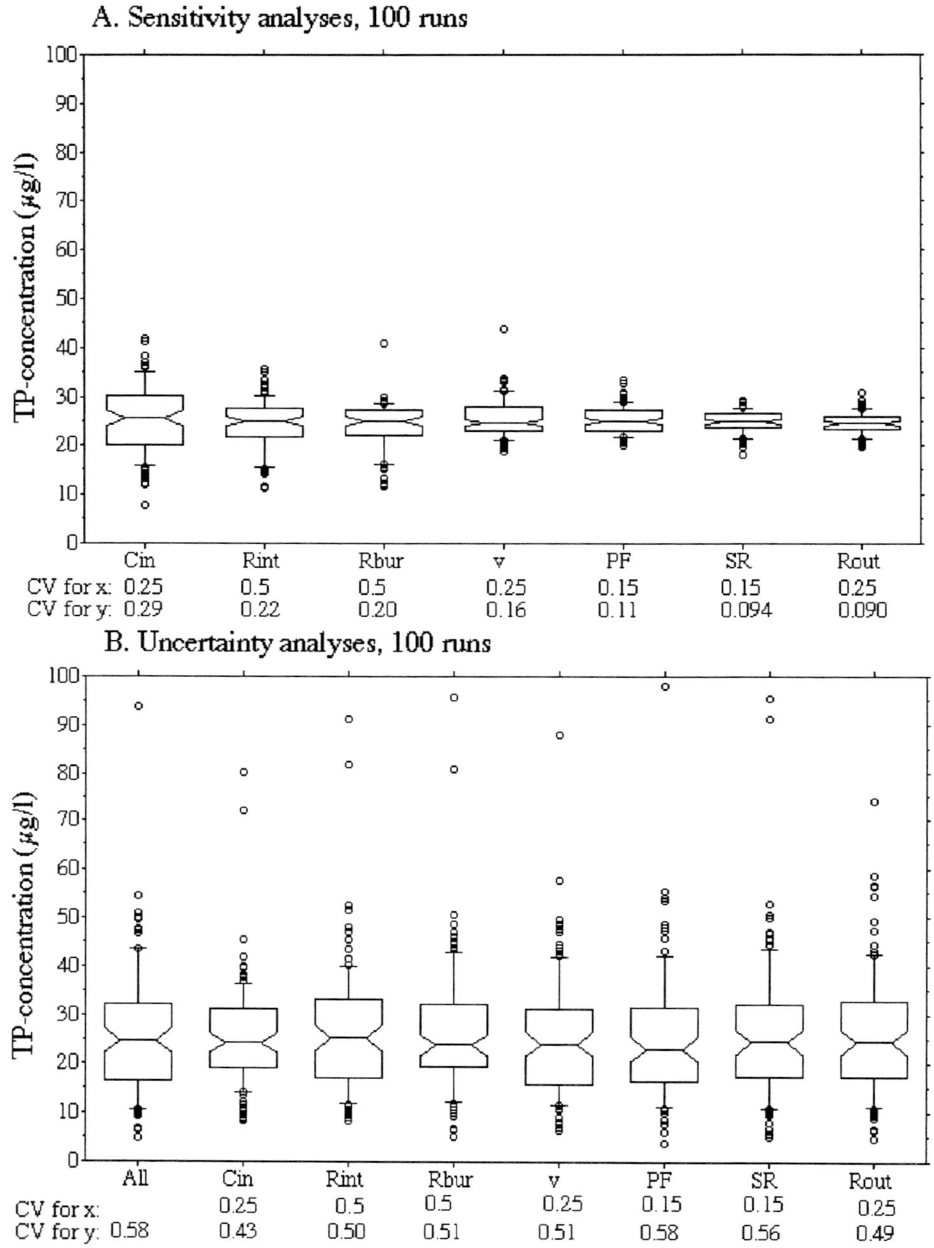

Figure 19. A. Sensitivity analyses using the mass-balance model and characteristic CV-values for seven x-variables for the target y-variable, lake TP-concentrations in Lake S. Bullaren. 100 runs. The box-and-whisker plots show the median (M50), quartiles, percentiles and outliers. The CV-values for the x-variables are given as well as calculated CV-values for the TP-concentration. B. Corresponding results from uncertainty analyses. All means that the uncertainties for all the given x-variables are included in the calculation of CV for y. The next results marked C_{in} means that the uncertainties related to C_{in} has been omitted and CV for y calculated by accounting for the remaining six uncertainties, etc.

Fig. 19 gives such an uncertainty analysis using the mass-balance model for lake concentrations of total phosphorus (see fig. 13; C_{TP}) and data from Lake S. Bullaren, Sweden. This figure also gives results (fig. 19A) for sensitivity analyses, which can be directly compared to the results for the uncertainty analyses (fig. 19B). One hundred runs have been done and the variations (including CV) for C_{TP} have been determined. One can note that the most crucial uncertainty component for C_{TP} also in this case is the value used for C_{in}, the tributary concentration. The values selected for the rate regulating internal loading (R_{int}), the rate regulating burial (R_{bur}), settling velocity (v), the particulate fraction (PF), the outflow or lake water retention rate (R_{out}) or the specific runoff rate (SR) influence TP predictions less. It is not shown here, but it can easily be demonstrated that the uncertainties associated with the morphometric data do not affect the predictions in any significant manner. The figure also gives the calculated CV for the target variable, C_{TP}.

Note that the ranking is the same for the most important variables influencing the target y-variable in fig. 19 (i.e., for C_{in}, R_{int}), but not for the following variables. This is because internal loading depends on outflow and sedimentation and such interrelationships are handled in a more comprehensive manner by uncertainty analyses than by sensitivity analyses, which only account for how one x-variable at the time influences the y-variable.

The results given in fig. 19 are indicative of a rather well balanced model. The calculated uncertainty for the target variable C_{TP} is also given. In more poorly balanced models, e.g., models of the Vollenweider- and OECD-type which do not account for internal loading and burial, the predictions of y are generally highly dependent on the reliability of one or a few x-variables, such as C_{in}. This is also the reason why so much effort (see Dillon and Rigler, 1974, 1975; Nichols and Dillon, 1978; Chapra and Reckhow, 1979, 1983) has been devoted to developing a good understanding of the processes regulating the tributary flux of phosphorus and the efforts to identify sources for TP-fluxes from catchments. A well-balanced model should not be too dependent on the uncertainty related to a single variable. Instead, all the box-and-whisker plots should look alike and the CV-values for y should not change too much if a model variable is omitted in the uncertainty analysis.

From these uncertainty tests, it is clear that the value for the tributary concentration, C_{in}, is very important in lake management and that uncertainty analysis is an essential tool to critically test the behaviour of mass-balance models.

DYNAMIC MODELS

The availability and treatment of empirical data becomes even more critical in dynamic modelling. To correctly describe dynamic processes in ecosystems does not only require a proper formulation of the physical processes, but also efficient ways of structuring the empirical data and proper simplifications to obtain time and area compatibility. Scales in time and space are discussed in the next two sections, and in the following sections water mixing and outflow in lakes will serve as examples of possible ways of describing complicated processes in a simple but efficient way.

Time Scale

The aim of the study must eventually determine the time scale of the model. However, there are inherent properties of uncertainty and accessibility at different scales that should influence the choice of time resolution. On one hand, the uncertainty – for example expressed by the coefficient of variation – of empirical data is smaller at smaller time scales, as illustrated in figure 8. On the other hand, it is nearly impossible or very laborious to access reliable data on a long term basis at short time scales, which means that there is a practical lower limit on the time resolution. There should be an optimal time scale where the combined shortcomings of uncertainty and accessibility have a minimum. This is illustrated in figure 20. In practical applications in contexts of lake studies, this optimum is generally found between one week and one month. In many situations, including modelling of emission scenarios and/or remedial measures, this is the required time scale.

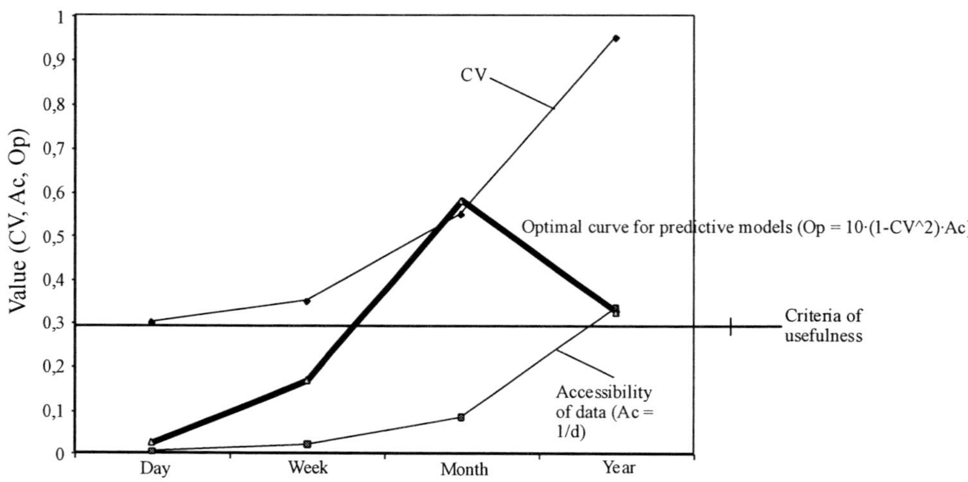

Figure 20. Illustration of factors regulating the optimal size of practically useful predictive models in ecology and radioecology

The Depth Separating Surface Water from Deep Water in Lakes

The question of space resolution is exemplified in this section. Apparently, there are several possible ways to deal with depth profiles of, e.g., temperature, substances and light availability for living organisms in lakes. The importance of temperature stratification is well known to all limnologists. In figure 21 temperature data from a Swedish lake are given for different time periods. For single sites after longer periods without strong winds, one can certainly see clear thermoclines, like in the summer of 1997 in the figure. However, for many sites over longer periods of time, it is apparent that the depth of the thermocline is extremely variable. Moreover, the depth of the thermocline may be very different at different sites within a lake, depending on local water currents and wind exposure. The same will be true also for concentrations of important chemicals and suspended particulate matter that governs the light availability for living organisms, but it does not mean that these depth profiles can be neglected. Rather, it means that one has to find effective ways to handle these problems in order to optimize the predictive power of the model. The space resolution of a model should not be finer than can be supported by observations at the chosen time scale. If the time scale is weeks or months, it should be possible to find a critical depth (D_{crit}, in meters) that would capture most of the important properties concerning profiles of temperature and substance concentrations. This is the operational water depth that differentiates the surface water (epilimnion) from the deep water (hypolimnion). The assumption that these two water masses can be treated as homogenous compartments can mostly be justified at a time scale chosen in such a manner. As shown in figure 21, the temperature profiles do not provide a consistent way to define these compartments. It has been shown that the depth separating areas where resuspension appears (erosion and transportation areas) from areas where continuous sedimentation is likely to prevail (accumulation areas) provides a useful estimation of the required critical depth (Håkanson et al., 2004). This depth may be calculated as:

$$D_{crit} = (45.7\sqrt{Area}) / (21.4+\sqrt{Area}) \qquad (6)$$

where Area is the lake area in m^2.

The volume of the surface water is given by the following equation (see Figure 22):

$$V_{SW} = V - V_{DW} = Area \cdot D_m - Area_A \cdot (D_{max} - D_{crit}/3 \qquad (7)$$

where D_m is the mean depth and D_{max} is the maximum depth in meters. $Area_A$ is the area below the wave base (the accumulation areas) and is calculated from the hypsographic curve as:

$$Area_A = ((D_{max} - D_{crit})/(D_{max} + D_{crit} exp(3 - Vd^{1.5})))\text{^}(0.5/Vd) \qquad (8)$$

The volume of surface water (V_{SW}) is simply the difference between the total volume (V) and V_{DW}

Water Mixing

Once the time scale and the depth separating surface water from deep water have been specified, it is possible to define methods to describe and predict processes of, e.g., water mixing and outflow. The proposed models predict the monthly rates of mixing and outflow for a given substance, but the time step may easily be altered.

The monthly mixing rate of a substance (R_{mix}; $month^{-1}$) may be described by a function of the absolute difference between mean monthly surface temperature (SWT) and deep water temperature (DWT).

$$\text{If ABS(SWT} - \text{DWT)} < 4° \text{ then } R_{mix} = 1 \text{ else}$$

$$R_{mix} = 1/ABS(SWT - DWT)) \qquad (9)$$

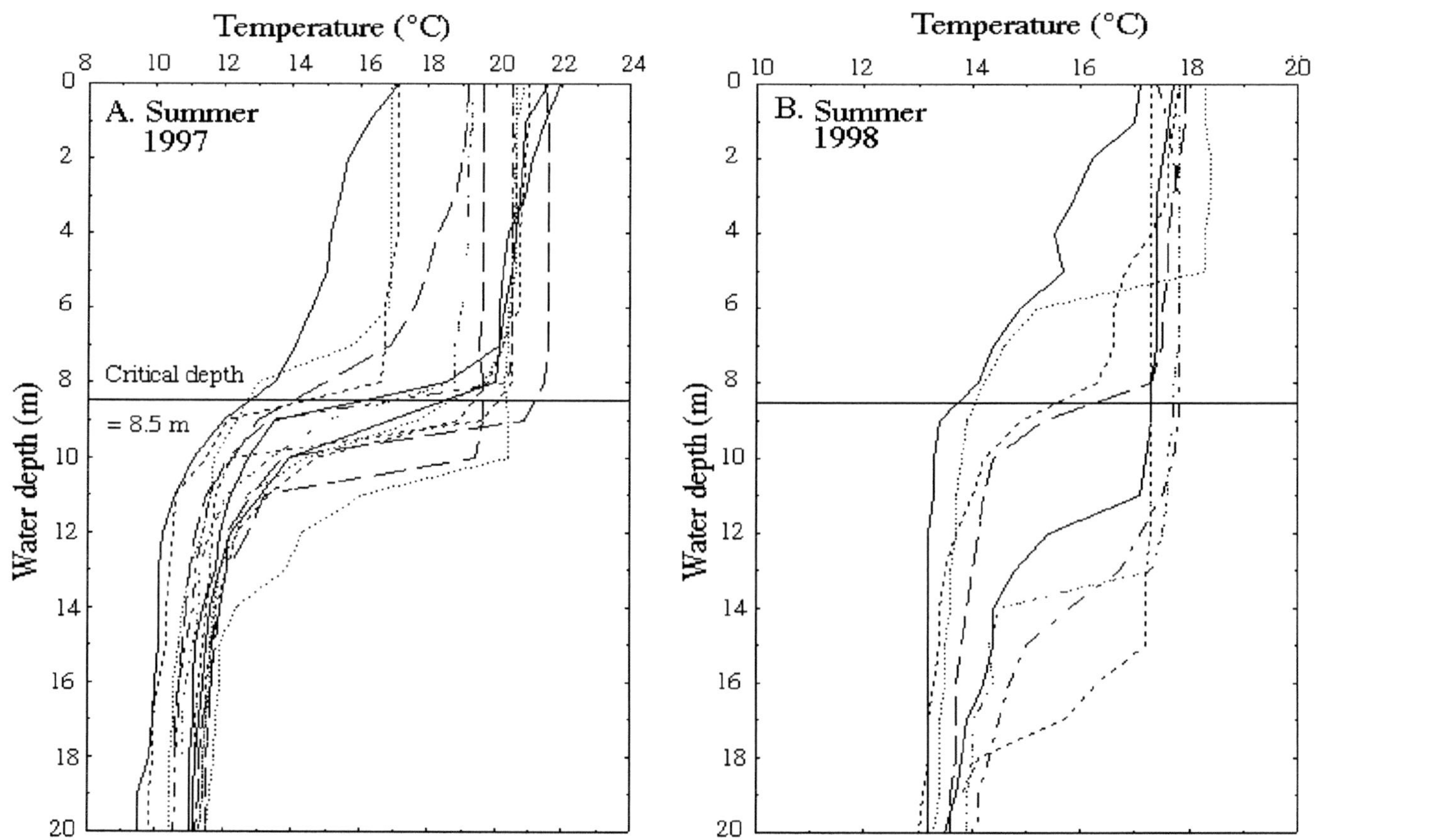

Figure 21. Temperature data from Lake Erken, Sweden (unpublished data for the summers of 1997 and 1998 from ongoing monitoring) and illustration of the critical depth (Dcrit, as defined in this paper) in this lake. Each line represents the temperature profile at the monitoring station at a given sampling occasion

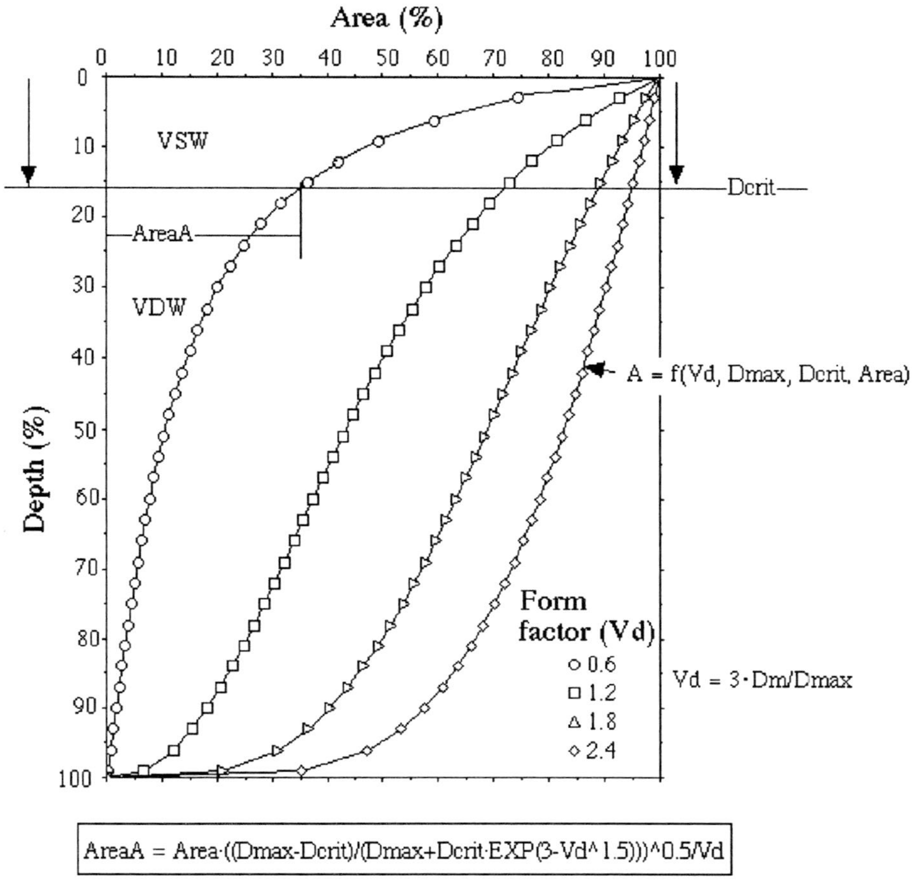

Figure 22. Illustration of relative hyposographic curves for four lakes with different forms (and form factor = volume development = Vd). The form influences the ET-areas above the critical water depth (Dcrit), and the A-areas below the critical water depth. The form of the lake is also important in defining the surface water volume (V_{SW}) and the deep water volume (V_{DW}).

Data on surface and deep water temperatures may, e.g., be calculated using the temperature model presented by Ottosson and Abrahamsson (1998). The mixing rate is never permitted to be higher than 1. The mixing from surface water to deep water will cause a transport given by:

$$F_{SWDWx} = M_{SW} \cdot R_{mix} \tag{10}$$

M_{SW} is the amount (in g) of the substance in the surface water compartment. Mixing from deep water to surface water is given by:

$$F_{DWSW} = M_{DW} \cdot R_{mix} \cdot V_{SW}/V_{DW} \tag{11}$$

Where M_{DW} is the amount (in g) of the substance in the deep water compartment. In the latter equation the equation is corrected for the difference in water volume between the deep water and surface water, implying that the same amount of water is transported in both directions.

Outflow

Generally, the outflow of a substance from a lake is given by:

$$F_{out} = Q_{out} \cdot C_{SW} \quad (12)$$

where F_{out} is the outflow of the given substance in g X per month. Q_{out} is the water transport out of the lake in m^3 per month. C_{SW} is the concentration of the substance in the surface water, given by:

$$C_{SW} = M_{SW}/V_{SW} \quad (13)$$

In this modelling approach, the balance between the precipitation of water on the lake surface and the evaporation are accounted for in a simple manner by two dimensionless moderators, Y_{SWT} and Y_{prec}. The water transport out of the lake (m^3 per month) is calculated as:

$$Q_{out} = Q_{in} \cdot Y_{SWT} \cdot Y_{prec} \quad (14)$$

Q_{in} is the water transport to the lake, which may be calculated from specific runoff or from the lake water retention time. The dimensionless moderator for surface water temperature is defined as:

If SWT < 9°C then $Y_{SWT} = 1$ else

$$Y_{SWT} = 1 - 0.4 \cdot (SWT/9 - 1) \quad (15)$$

where 9 is the norm-temperature related to the duration of the growing season (Håkanson and Boulion, 2002) and 0.4 is the amplitude value. This means that $Y_{SWT} = 1$ if the actual value of the mean monthly surface water temperature is 9°C; Y_{SWT} will approach zero when the actual mean monthly surface water temperature approaches 33°C. Then the outflow of water (Q_{out}) also approaches zero. The dimensionless moderator for precipitation (Y_{prec}) is defined by:

If Prec < 650 mm per year then

$Y_{prec} = (1 + 1.8 \cdot (Prec/650 - 1))$ else

$$Y_{prec} = (1 + 0.5 \cdot (Prec/650 - 1)) \quad (16)$$

Where 650 is the norm for the mean annual precipitation (Prec); 1.8 is the amplitude value if Prec is smaller than 650; 0.5 is the amplitude value if Prec is larger than 650 mm per year. This means that $Y_{prec} = 1$ if Prec =650; Y_{prec} will approach zero when the actual mean annual precipitation approaches 300 mm per year, and the outflow of water (Q_{out}) also approaches zero; Y_{prec} will increase when the actual mean annual precipitation increases. The outflow of water (Q_{out}) is twice as large when Prec is 2000 mm per year compared to when Prec = 650 mm per year. The presented model for outflow has been tested with good results by Håkanson et al. (2004).

REFERENCES

Beck, M.B. and Van Straten, G., 1983 (eds). *Uncertainty, system identification and the prediction of water quality.* Springer, Heidelberg, 387 p.

Bodemer, U., 1998. DF-excitation spectroscopy of phytoplankton: relationship between dynamics of algal populations and discharge. Large Rivers Vol. 11, No. 2. *Arch. Hydrobiol. Suppl.*, 115/2:125-138.

Chapra, S.C. and Reckhow, K., 1979. Expressing the phosphorus loading concept in probabalistic terms. *J. Fish. Res. Bd. Can.*, 36:225-229.

Chapra, S.C. and Reckhow, K., 1983. *Engineering approaches for lake management. Vol. 2. Mechanistic modelling.* Butterworth, Woburn, Mass.

Cox, D.C. and Baybutt, P., 1981. Methods for uncertainty analysis: a comparative survey. *Risk Analysis*, 1: 251-258.

Dillon, P.J. and Riegler, F.H., 1974. A test of a simple nutrient budget model predicting the phosphorus concentration in lake water. *J. Fish. Res. Board Can.*, 31:1771-1778.

Dillon, P.J. and Rigler, F.H., 1975. A simple method for predicting the capacity of a lake for development based on lake trophic status. *J. Fish. Res. Board Can.*, 32:1519771-1531.

Foster, I.D.L., Baban, S.M.J., Wade, S.D., Charlesworth, S., Buckland, P.J. and Wagstaff, K., 1996. Sediment-associated phosphorus transport in the Warwickshire River Avon, UK. IAHS Publ. (Proc. Exeter Symposium), 236:303-312.

Foster, I.D.L., Baban, S.M.J., Charlesworth, S.M. Jackson, R., Wade, S., Buckland, P.J., Wagstaff, K. and Harrison, S., 1997. Nutrient concentrations and planktonic biomass (chlorophyll *a*) behaviour in the catchment of the river Avon, Warwickshire *IAHS Publ. (Proc. Rabat Symposium),* 243:167-176.

Foster, I.D.L., Wade, S., Sheasby, J., Baban, S., Charlesworth, S., Jackson, R. and McIlroy, D., 1998. Eutrophication in controlled waters in Lower Severn Area. *Environment Agency, Severn-Trent Region* (Final Report), Vol. 1, 94 pp., Vol. 2, 163 pp., Tewkesbury, UK.

Friedrich, G., Gerhardt, V., Bodemer, U. and Pohlman, M., 1998. Phytoplankton composition and chlorophyll concentration in freshwater: comparison of delayed fluorescence excitation spectroscopy, extractive spectrophotometric method and Utermöhl-method. *Limnologica,* 28:323-328.

Gilbert, R.O., 1987. *Statistical methods for environmental pollution monitoring.* Van Nostrand Reinold Co., New York.

Håkanson, L., 1999. *Water Pollution - methods and criteria to rank, model and remedi-ate chemical threats to aquatic ecosystems.* Backhuys Publishers, Leiden, 277 p.

Håkanson, L., Blenckner, T. and Malmaeus, J.M., 2004. Defining new, general methods to define surface water and deep water, outflow and internal loading for mass-balance model for lakes. – *Ecol. Modelling*, 175:339-352.

Håkanson, L. and Boulion, V., 2002. *The Lake Foodweb - modelling predation and abiotic/biotic interactions.* Backhuys Publishers, Leiden, 344 p.

Håkanson, L. and Jansson, M., 1983. *Principles of Lake Sedimentology.* Springer, Berlin, 316 p.

Håkanson, L., Malmaeus, J.M., Bodemar, U. and Gerhardt, V., 2003. Coefficients of variation for chlorophyll, green algae, diatoms, cryptophytes and blue-greens in rivers as a basis for predictive modelling and aquatic management. *Ecol. Modelling*, 169:179-196.

Håkanson, L. and Peters, R.H., 1995. *Predictive limnology- methods for predictive modelling.* SPA Academic Publishers, Amsterdam, 464 p.

Hamby, D.M., 1995. A comparison of sensitivity analysis techniques. *Health Physics*, 68:195-204.

Hinton, T. G., 1993. Sensitivity analysis of ecosys-87: an emphasis on the ingestion pathway as a function of radionuclide and type of disposition. *Health Physics*, 66:513-531.

IAEA, 1998. International Atomic Energy Agency. Modelling of radiocesium in lakes. *Tec. Doc.* (in print), Vienna.

Nicholls, K.H. and Dillon, P.J., 1978. An evaluation of phosphorus-chlorophyll-phytoplankton relationships for lakes. *Int. Revue ges. Hydrobiol.*, 63:141-154.

OECD, 1982. *Eutrophication of waters. Monitoring, assessment and control.* OECD, Paris, 154 p.

Ottosson, F. and Abrahamsson, O., 1998. Presentation and analysis of a model simulating epilimnetic and hypolimnetic temperatures in lakes. *Ecol. Modelling*, 110:223-253.

Pilesjö, P., Persson, J. and Håkanson, L., 1991. Digital bathymetric information for calculations of coast morphometrical parameters and surface water retention time (in Swedish with English summary). *Swedish Environmental Protection Agency, Report* 3916, Stockholm, 76 p.

Rose, K.A., McLean, R.I. and Summers, J.K., 1989. Development and Monte Carlo analysis of an oyster bioaccumulation model applied to biomonitoroing. *Ecol. Mod.*, 45:111-132.

Swale, E.M.F., 1964. A study of the phytoplankton of a calcareous river. *J. Ecol.* 52:433-445.

Tiwari, J.L. and Hobbie, J.E., 1976. Random differential equations as models of ecosystems. Monte Carlo simulation approach. *Math. Biosci.*, 28:25-44.

Vollenweider, R.A., 1968. The scientific basis of lake eutrophication, with particular reference to phosphorus and nitrogen as eutrophication factors. *Tech. Rep. DAS/DSI/68.27, OECD,* Paris, 159 pp.

Vollenweider, R.A. and Kerekes, J., 1980. OECD cooperative programme for monitoring of inland waters. Eutrophication control. *Synthesis Report, OECD*, Paris.

Whicker, F.W., 1997. Measurement quantities and units. *Preprint from an International Workshop on measuring radionuclides in the environment: radiological quantities and sampling designs*, Bad Honnef, Germany, Nov., 1997.

Worley, B.A., 1987. Deterministic uncertainty analysis. *Oak Ridge National Laboratory Report ORNL-6428*, Oak Ridge, U.S.A., 53 p.

In: New Trends in Ecology Research
Editor: A. R. Burk, pp. 71-102
ISBN: 1-59454-379-8

Chapter 3

EFFECT SIZE ESTIMATES AND CONFIDENCE INTERVALS: AN ALTERNATIVE FOCUS FOR THE PRESENTATION AND INTERPRETATION OF ECOLOGICAL DATA

Julian Di Stefano,[1] Fiona Fidler[2] and Geoff Cumming[3]
[1]School of Forest and Ecosystem Science and Department of Zoology, University of Melbourne
[2]School of Botany and Department of History and Philosophy of Science, University of Melbourne
[3]School of Psychological Science, La Trobe University

ABSTRACT

Since the development and popularisation of statistical science during the 1930s, the uptake of statistical hypothesis tests in major science disciplines like psychology, medicine and biology has been rapid. Hypothesis tests did not become common in the ecological literature until the 1960s but, since then, testing zero null hypotheses and using P-values to make dichotomous decisions about rejecting them has become routine. Recent surveys of the ecological literature show that P-values are reported in the vast majority of published research articles.

We believe that the focus on tests of zero null hypotheses and the reliance on P-values is inappropriate and argue that ecological data are more relevant to both ecologists and applied users if *a priori* sample size determination procedures are conducted, and effect size estimates and their associated confidence intervals are used to present results. We report evidence that effect sizes, associated uncertainty estimates and sample size calculations are missing from a large number of publications in two leading conservation biology journals. An overwhelming majority (92%) of surveyed articles reported P-values, but only 3% of these reported statistical power. Furthermore, many articles were missing at least one estimate of effect size (43%) or measure of variance (67%). Confidence intervals were reported in only 19% of articles and, when reported, were often inadequately interpreted.

An ecological data set is analysed using both P-values and confidence intervals, and the results from these two approached compared. We also demonstrate a novel precision-based approach for *a priori* sample size calculations that is consistent with the philosophy of interval estimation. Although we advocate the use of confidence intervals, recent research shows that they are often incorrectly interpreted. Two major problems are (a) under estimation of the degree to which two confidence intervals must overlap before their associated means are statistically different and (b) overestimation of the proportion of future means enclosed by a sample interval. We provide guidelines for the interpretation and presentation of confidence intervals so that users of ecological data can make the most of the information they contain.

INTRODUCTION

In this chapter, our primary objective is to present a sensible strategy for the analysis and presentation of ecological data. We believe that, in the majority of cases, the quality and accessibility of research outputs will be improved by (a) conducting *a priori* sample size determination procedures and (b) presenting results using ecologically relevant effect size estimates and associated measures of uncertainty, a process commonly referred to as interval estimation. Proponents of interval estimation are often highly critical of traditional hypothesis testing practices, but we feel it is important to note that interval estimation and hypothesis testing are not mutually exclusive. In fact many ecological studies have as their primary aim comparisons between two or more parameter estimates or between a parameter estimate and an ecologically relevant standard, and as such can be analyses in relation to a null and alternative hypothesis. What we regard as an uninformative practice is testing zero null hypothesis (nulls do not have to be hypotheses of no difference) and the use of P-values to make dichotomous decisions about study results.

Although the analyses we describe operate within a frequentist (or classical)[1] statistical framework, we are not married to a particular data analysis paradigm. In the ecological literature there is growing support for both Bayesian statistics (Ellison 1996, 2004; Wade 2000) and model selection approaches (Anderson *et al.* 2000; Burnham and Anderson 1998, 2001), and enhanced graphical presentation of data has also been advocated for some time (Tukey 1977). There are often a number of rational ways to analyse an ecological data set, and Bayesian, model selection or simple graphical analysis are all useful in particular contexts. Our aim is to describe an alternative that is relevant in many situations.

Before beginning our discussion about hypothesis testing, P-values and confidence intervals, we present a brief historical sketch outlining the development of classical statistical inference throughout the first half of the twentieth century, the rapid uptake of null hypothesis testing by a number of disciplines in the years after World War Two, and the more recent debate concerning the utility of this practice. This is presented in section 2. This history has been well documented by others (Fidler *et al.* in press; Gigerenzer 1993; Gigerenzer *et al.* 1989; Senn 1997), and we rely heavily on these sources. Nevertheless, we feel a brief retelling is appropriate for a number of reasons. First, most ecologists are unaware of the historical developments associated with classical statistics and the hypothesis testing debate,

[1] Throughout this chapter we use the terms 'frequentist' and 'classical' interchangeably to describe the branch of statistics that draws inferences by comparing a test statistic generated from sample data to the theoretical probability distribution for that statistic. This comparison is used to generate probability statements that are correctly interpreted as a long-run frequency, although many users of frequentist statistical techniques do not interpret the output in this way.

and we feel it is an interesting and informative story. More importantly, the history outlines the parallel existence of two different and conflicting analytical approaches and their unreconciled merger after World War Two, and helps to explain many of the problems associated with null hypothesis testing and the interpretation of P-values that exist today. It also reveals the move away from null hypothesis testing by some other disciplines, and strengthens the legitimacy of alternative approaches.

In section 3 we identify null hypothesis testing and the use of P-values as the predominant current framework for the analysis and presentation of ecological data, outline criticisms of this approach, and present survey data that demonstrate the current poor state of statistical reporting in conservation biology journals. In section 4 we describe how effect size estimation and error presentation using confidence intervals can provide more information than the traditional hypothesis testing approach. Examples are presented that illustrate the use of confidence intervals for both data interpretation and sample size determination. In section 5 we outline some of the challenges confronting interval estimation, and present some preliminary data exploring the way confidence intervals are understood and interpreted by both students and established researchers. We conclude with a call for ecologists to consider alternatives to zero null hypotheses, for research examining cognitive responses to confidence intervals, and for journal editors to promote improved methods for the analysis and interpretation of ecological data.

A Brief History of Classical Statistical Inference and the Hypothesis Testing Debate

The development of classical statistical science is often attributed to Sir Ronald Fisher who produced a number of influential texts throughout the first half of the 1900s (Fisher 1925, 1935, 1956). Modern classical statistics, however, is really a blend of Fisher's ideas with those of Jerzy Neyman and Egon Pearson (Neyman 1950, 1955, 1957; Neyman and Pearson 1928; Pearson 1962). Neyman and Pearson originally developed their theory to complement and extend Fisher's work, but a series of intellectual and personal disagreements between Fisher on the one hand and Neyman and Pearson on the other resulted in an often acrimonious discourse that lasted from the 1930s until Fisher's death in 1962. Although the debate between the parties ended at this time, the conflict between the theories remains (Gigerenzer 1993).

The inferential framework proposed by Fisher advocated the specification and testing of null hypotheses (Fisher 1935), the modern derivation of which has become known as null hypothesis significance testing, or NHST. The procedures were based on frequentist principles and involved specifying a null hypothesis and then exposing this null to data. If the data were sufficiently departed from the null, the null was said to be implausible. The degree to which the data and the null had to be separated before the null was considered implausible was defined by Fisher as the 'level of significance' and computed (using significance tests such as the *t*-test) as P(D/H), the probability of obtaining the data (or data more extreme) assuming that the null hypothesis was true (such outputs are commonly called P-values, and shall be referred to as such henceforth). This is in stark contrast to Bayes Theorem (Bayes 1763), another prominent inferential technique, which enables the calculation of P(H/D) – the probability of a hypothesis given a particular set of data. This is often what scientists really want to know (Cohen 1994) and can lead to the misinterpretation of P-values, an issue that will be covered in more detail later.

If the P-value resulting from a significance test was low (Fisher suggested 0.05 or 0.01 as conventions), the null hypothesis was considered sufficiently implausible. Fisher considered significant results (ie, P-values below the pre-specified cut off point) to be evidence directly related to the specific null hypothesis being tested (a view that, strictly speaking, is inconsistent with the frequentist precept upon which his inferential framework was based), but believed that non-significant results did not represent evidence one way or the other and should be ignored (Fisher 1935). His framework was asymmetrical in the sense that it only specified a single statistical hypothesis (the null), and only incorporated Type I errors (rejecting the null hypothesis when it is actually true in the real world). Alternative hypotheses, Type II errors (failing to reject the null when it is in fact false) and statistical power (1 minus the Type II error rate) were rejected by Fisher as inappropriate for scientific inference and discovery (Gigerenzer 1993).[2]

Like Fisher, Neyman and Pearson believed that inferential procedures should operate within a frequentist framework, but took a more literal approach to the interpretation of frequentist statistical outputs. While Fisher believed that the results of a significance test were directly related to a particular experiment, Neyman-Pearson theory interpreted P-values as the probability of a Type I error over the long term. For Neyman and Pearson, a P-value did not relate to the sample of data used to generate it but to a notional infinite sequence of samples taken from the same population (Neyman 1950).

Neyman-Pearson theory also specified statistical null and alternative hypotheses, one of which was assumed to be true (Neyman and Pearson 1928). This altered the interpretation of non-significant results from 'inconclusive' in Fisher's case to evidence supporting the null hypothesis. The formal specification of two statistical hypotheses also enabled the probability of both Type I and Type II errors to be calculated, and Neyman and Pearson referred to these two values as α (alpha) and β (beta). The formal definition of β meant that the power of a statistical test could be derived, a concept defined as the long run probability of rejecting the null hypothesis (ie, getting a statistically significant result) if it is in fact false in the real world. With two competing hypotheses and a formal definition of both α and β, Neyman-Pearson theory advocated *a priori* cost-benefit analysis to define appropriate Type I and Type II error rates for a particular experimental scenario, and thus paved the way for modern *a priori* power calculations. Once these *a priori* considerations had been made, however, the alpha level at which to reject the null hypotheses remained fixed and in this sense, Neyman and Pearson were strong proponents of the dichotomous decision to accept or reject the null hypothesis depending on which side of the line the P-value fell. Although this is consistent with Fisher's early work (Fisher 1935), later in his life he changed his mind and criticised this approach as far too restrictive (Fisher 1955); see footnote 2).

There are clearly a number of irreconcilable differences between Fisher's version of frequentist inference and Neyman-Pearson theory. Fisher wan not concerned with alternative hypotheses, Type II error rates or sample size calculations, and (at least in his later work)

---

[2] Fisher's inferential logic was somewhat inconsistent and substantial differences on a number of key points can be found between his earlier and later works. For a detailed account of the development of Fisher's logic and the changes that occurred throughout his life see Gigerenzer (1993) and Gigerenzer et al. (1989).

interpreted P-values as evidence for or against a particular hypothesis. Neyman-Pearson theory, on the other hand, specified prior definition of alternative hypotheses and significance levels, and encouraged the calculation of both Type I and Type II error rates. Neyman and Pearson took frequentist inference literally, and interpreted P-values as the long run probability of a Type I error.

In the years immediately following World War Two formalised statistical procedures became increasingly popular in some fields of experimental science, particularly psychology, medicine and associated sub-disciplines (Fidler *et al.* in press; Hubbard and Ryan 2000). This was greatly facilitated by a number of new textbooks written specifically for practitioners that advocated null hypothesis testing. Many of these texts, however, attributed the development of hypothesis testing to Fisher but combined Fisher's ideas with Neyman-Pearson concepts without acknowledging their original source or discussing the different theoretical and somewhat conflicting mathematical and epistemological frameworks. For example, P-values were often described as evidence for or against a particular hypothesis (Fisher) but Type I and Type II errors explained as a long run probability (Neyman-Pearson). This difference, compounded by the obvious confusion of some of the authors, lead to conflicting messages about the meaning of P-values that worked their way into the consciousness of practising scientists (Gigerenzer 1993).

The net result within the psychological and medical scientific communities was the adoption of a mechanical, apparently unified data analytic framework advocating null hypothesis tests and the use of P-values as the predominant method for presenting results. Although this practice did not become prominent within the ecological community until the late 1960s, since then the uptake of null hypothesis testing has been rapid and adopted as the analytical method of choice in a number of influential books and papers (Downes *et al.* 2002; Quinn and Keough 2002; Sokal and Rohlf 1995; Underwood 1990; Zar 1999). This, combined with the development of easy to use hypothesis testing software and a publication environment biased towards statistically significant results (Palmer 2000), has made null hypothesis tests and P-values the most widely used method of analysing and reporting ecological data.

To be fair, modern statistical texts aimed at the ecological community do warn against some of the problems associated with null hypothesis testing. For example, (Quinn and Keough 2002) include a section outlining the arguments against hypothesis testing, discuss Bayesian analysis and encourage the presentation of effect sizes and measures of uncertainty, while (Sokal and Rohlf 1995) suggest scientists use terms such as 'important' and 'meaningful' to differentiate between statistical significance and biological importance.[3] As the data presented in the next section show, however, these recommendations are often not incorporated into published ecological studies.

Most ecologists are probably unaware of the major debate within psychology, medicine and related fields regarding the use and interpretation of hypothesis tests and P-values. The American Psychological Association, for example, convened a taskforce whose objective was '...to elucidate some of the controversial issues surrounding applications of statistics including significance testing and its alternatives...' (Wilkinson 1999). The report contained

---

[3] Although Sokal and Rohlf (1995) differentiate between statistical significance and biological importance, throughout their book they 'accept null hypotheses' in response to P-values ≥0.05, an inappropriate and misleading conclusion in many cases.

many recommendations for the analysis and reporting of data, including a call for the presentation of effect sizes and an associated measure of uncertainty, advice that is repeated in the editorial policies of many medical journals. In some extreme cases (eg, *American Journal of Public Health* and *Epidemiology*), editors banned the use of P-values altogether (Fidler *et al.* 2004).

The debate is also active among statisticians (Chatfield 1985; Cox 1977; Nelder 1999; Sellke *et al.* 2001). Nelder (1999), for example, refers to the general overemphasis on P-values (the P-value culture, as he calls it) as non-scientific, and contends that the overuse of P-values obstructs the accumulation of scientific information. (Sellke *et al.* 2001) identify that P-values are often incorrectly interpreted and, in tests of precise null hypotheses, a P-value of 0.05 may not provide particularly strong evidence against the null.

A number of papers have recently appeared in the ecological literature criticising the way that null hypothesis tests and P-values are used (Anderson *et al.* 2000; Di Stefano 2004; Ellison 1996; Johnson 1999, 2002; Osenberg *et al.* 2002; Yoccoz 1991), and a number of these, and some others, have advocated interval estimation as a complementary or alternative strategy (Anderson *et al.* 2001; Di Stefano 2004; Gerard *et al.* 1998; Johnson 1999; Osenberg *et al.* 2002; Steidl *et al.* 1997; Steidl and Thomas 2001; Yoccoz 1991). We believe these issues deserve serious consideration, and we encourage ecologists to take part in the debate.

STATISTICAL INFERENCE IN ECOLOGY

Unlike psychology which embraced modern statistical procedures during the 1950s (Hubbard and Ryan 2000), statistical tests of null hypothesis did not become routine in the ecological literature until the second half of the 1960s (Fidler *et al.* in press). Since then the use of statistical tests, and the reporting of P-values in particular, has increased at a rapid rate. (Anderson *et al.* 2000) surveyed papers published in the highly regarded journal *Ecology* between 1978 and 1997 and found that the average number of P-values per paper rose from 10 - 20 in the late 1970s and early 1980s to 25 - 40 during the late 1980s and 1990s. Remarkably, one of the sampled papers from 1984 contained 317 P-values, and three others from 1991, 1996 and 1997 contained 204, 208 and 208 respectively. (Anderson *et al.* 2000) also surveyed papers from the *Journal of Wildlife Management*, and the results revealed a similar trend. A survey we conducted of 50 papers each from *Conservation Biology* and *Biological Conservation* from 2000 and 2001 issues showed that 92 out of the 100 reported P-values. Together with surveys of journals in other sub-fields (Peterman 1990; Thomas and Juanes 1996), these provide strong evidence that the use of classical statistical techniques and the presentation of P-values is prevalent in ecological research.

Although we believe that classical statistical procedures are frequently useful for planning experiments and analysing ecological data, there are a number of problems with the way these methods are currently used, interpreted and reported in the ecological literature. These problems have been discussed in detail elsewhere, and the main criticisms are outlined below.

Mechanistic Use of Statistical Procedures

In much of the published ecological literature, statistical procedures are used mechanistically and are frequently viewed as objective mathematical rules that provide unambiguous and rigorous answers to scientific questions. In reality, however, appropriate use of statistical procedures and the interpretation of their outputs involves judgements informed by the theoretical, physical and social context surrounding each study (Stewart-Oaten 1995). The mechanistic use of statistics is a modern phenomenon and was not advocated by Fisher, Neyman or Pearson; a number of references can be found in their work stating the importance of judgement in statistical inference (Gigerenzer 1993).

Testing Uninformative Null Hypotheses

Commonly, statistical null hypotheses state that there is no difference between parameters, or that a parameter equals zero, propositions that are almost always false. It is well documented that the use of statistical tests to reject such nulls is generally trivial and uninformative (Anderson et al. 2000; Johnson 1999). As an example, consider an experiment designed to test the general theory that mammals living at sites with high resource quality will have smaller home ranges than the same species occupying sites with low quality resources. To test this theory, individuals are trapped at both high and low quality sites, and their home ranges recorded. Traditionally, the null hypothesis is specified as H_0: $\mu_1 = \mu_2$, the mean home range size at high quality sites equals the mean home range size at low quality sites. This null is trivial because, regardless of the impact of resource quality on home range size, the mean of one group is almost certain to differ from the mean of the other. It is uninformative because its rejection is inevitable given adequate sample size. Important ecological information, such as the mean home range size of animals living at high and low quality sites, the difference between these means, and the errors associated with these parameter estimates, are not related to the specification of a zero null hypothesis.

Over Reliance on P-Values

Our survey (presented in detail later), and the survey conducted by (Anderson et al. 2000), show that P-values are used frequently in the ecological literature, and often presented as a results summary without additional information such as effect sizes or error estimates.

Uninformative Nature of P-Values

In many ecological experiments, estimates of effect size and the errors associated with these estimates are the most important outputs from the data (Di Stefano 2004; Johnson 1999). Consider the classical (and very common) form of manipulative experiment where an experimenter imposes different states (treatments) on groups of replicated experimental units and records the effect by measuring a meaningful response variable. In this case, the parameters of interest are (a) the estimated treatment effects, that is, the change in the response variable resulting from each treatment, relative to other treatments or to a control condition, and (b) some measure of how accurate these estimates are. All P-values do is specify if an effect is likely (or unlikely) to exist. They provide no information about the size of an effect or its precision.

Misinterpretation of P-Values

As stated earlier, the P-value is the probability of obtaining data (or data more extreme), assuming that the null hypothesis is true. Put another way, a P-value specifies how consistent (or inconsistent) a data set is with the null hypothesis (Ellison 1996). Nevertheless, many scientists interpret P-values as something different. A number of common misinterpretations are:

- P is probability that observed effects are due to chance.
- 1-P is the reliability of a result (ie, the probability of obtaining a similar result if the study were repeated).
- P is the probability that the null hypothesis is true (and therefore 1-P is the probability that the alternative hypothesis is true). This is the correct interpretation of a probability statement derived from Bayesian analysis, and is really what most researchers want to know. In fact P-values calculated using classical frequentist approaches can be '...highly misleading measures of the evidence provided by the data against the null hypothesis.' (Berger and Sellke 1987).
- P-values are indicative of effect sizes (ie, small P-values mean large effects).
- P-values are indicative of ecological importance (ie, small P-values mean ecologically important effects). (Yoccoz 1991) provides an eloquent summary of the difference between statistical significance and ecological importance.

Although some ecologists do not interpret P-values incorrectly, many do (Ellison 2004), and these interpretive errors can lead to misinformed and unfounded conclusions.

Use of Arbitrary Decision Criteria

In his book about the use of Analysis of Variance in ecology, (Underwood 1997) has written:

> Biologists...have been obsessed with conventional and arbitrary views about the probability of Type I error, so that we use P = 0.05 as though it were written on a tablet of stone by some god of statistical theory.

There is no god of statistical theory and a Type I error rate of 5% is not sacred. In fact the notion of specifying a cut-off point below which results are not significant and above which they are is ecologically uninformative (Anderson *et al.* 2000; Johnson 1999). Unfortunately this is still an all too common occurrence in the ecological literature - a browse through a recent issue of *Austral Ecology* at the time of writing revealed the specification of P = 0.049 as significant and P = 0.059 as not significant (Barrick 2003). As (Rosnow and Rosenthal

1989) have said, 'surely God loves 0.06 nearly as much as 0.05'. The use of an arbitrary cut-off point (whether 0.05 or another value) to specify statistical significance has little ecological meaning and promotes the erroneous notion that every experiment supports an unambiguous conclusion (Smithson 2003).

Under Use of *A Priori* Sample Size Determination Techniques

Classical statistical techniques can be used in the planning phase of studies to estimate the sample size required to detect an effect of specified magnitude. In the ecological literature this has been most commonly discussed in the context of a priori power analysis (Di Stefano 2001; Downes et al. 2002; Fairweather 1991; Foster 2001; Keough and Mapstone 1997; Mapstone 1995), but confidence intervals can also be used. We demonstrate an application of this technique in Section 4.3.

Sample size calculations prior to an experiment are useful as researchers are forced to specify their statistical model and consider the size of ecologically important effects and their expected variances. In the case of *a priori* power analysis, the relative cost of Type I and Type II errors and how these two parameters interact must also be contemplated (Di Stefano 2003; Fairweather 1991; Mapstone 1995). Nevertheless, *a priori* sample size determination is infrequently conducted in ecology resulting in parameter estimates that are often imprecise and uninformative (Fairweather 1991; Mapstone 1995; Peterman 1990).

Statistical Reporting Practices in Conservation Biology and Biological Conservation

In order to substantiate claims that statistical reporting practices are often inadequate in the ecological literature, we conducted a detailed survey of 50 *Conservation Biology* and 50 *Biological Conservation* articles published in 2000 and 2001. We surveyed a systematic sample of articles that contained new empirical data, but excluded meta-analyses, methodological and theoretical articles. We recorded a number of variables (outlined below), and calculated the proportion of articles reporting each one, along with the 95% confidence intervals for those proportions using the method described by (Newcombe and Altman 2000). The variables we recorded and the results for each are outlined below:

Null Hypothesis Tests

Overall, 92% (92 of 100; 95% confidence interval: 85 to 96%) of articles tested statistical null hypothesis.

Null Hypothesis of No Difference

Of the 92% that tested null hypothesis, 79% (73 of 92; 70 to 86%) tested a null hypothesis of no difference. Due to insufficient information, this may be an underestimate.

Statistical Non-Significance and Statistical Power

Most articles (80%, 74 of 92; 71 to 87%) reported at least one statistically non-significant result and about half of these (47%, 35 of 74; 35 to 57%) interpreted the statistically non-significant result as evidence for "no effect" or "no relationship". However, only 3% (2 of 74;

0 to 9%) reported statistical power. A further 30% (22 of 74; 21 to 41%) made an implicit reference to power issues, such as noting that the sample size was small.

Ambiguous Use of 'Significant'

Two thirds of articles (68%, 63 of 92; 58 to 77%) used the word 'significant' ambiguously. If the author did not preface 'significant' with 'statistically', or follow it with a P-value or test statistic, or otherwise differentiate statistical and substantive interpretations, the practice was recorded as 'ambiguous'.

P-Value Reporting

One quarter (23 of 92; 17 to 35%) of articles that tested null hypothesis reported statistical significance using asterisks and/or or predefined cut-off levels (eg <0.05) while 62% (56 of 92; 51 to 70%) reported exact P-values. Use of asterisks (one, two, or three star significance) has been heavily criticised (Meehl 1978) as this practice provides even less information than exact P-values, is usually insufficient for meta-analysis and has the potential to mislead researchers into thinking that an effect with two stars is more important than an effect with one.

Effect Sizes, Variance Measures, and Sample Sizes

Most articles reported at least one effect size, variance measure, and sample size, but in many articles at least one of these items was missing. Summary data are presented in Table 1.

Confidence Intervals

Of the 100 articles, 19% (13 to 28%) reported confidence intervals, and 26% (5 of 19; 12 to 49%) attempted to interpret them. Any mention of a confidence interval beyond its mere reporting, was recorded as an interpretation.

Figures

A figure was any visual representation of new data and 77% of articles provided at least one. Of these 40% (31 of 77; 30 to 51%) included error bars, representing standard deviation (n=6), standard error (n=14) or confidence interval (n=9). In two cases, bars were unlabelled.

Table 1. Summary of reporting practices in a sample of 100 empirical articles from *Biological Conservation* and *Conservation Biology* with respect to effect size, standard deviation, standard error, sample size and confidence intervals

Reporting practice	% of articles	95% CI
Reporting at least one effect size	87 (80 of 92)	80-94
Missing at least one effect size	43 (40 of 92)	34-53
Reporting at least one SD or SE	48 (44 of 92)	38-58
Missing at least one SD or SE	67 (62 of 92)	58-77
Reporting at least one sample size or df	76 (70 of 92)	68-84
Missing at least one sample size or df	36 (33 of 92)	26-45
Reporting at least one CI	19 (19 of 100)	11-27
Interpreting CIs	26 (5 of 19)	3-39

In general, our survey shows that null hypothesis testing and P-values are the dominant analytical and data presentation method used in the *Conservation Biology* and *Biological*

Conservation, a result consistent with other surveys of the ecological literature. Specifically, the data indicate that there is a general methodological failure in these journals which requires urgent attention.

Of major concern is (a) the use of P-values alone without associated effect size estimates and measures of precision, (b) the interpretation of non-significant results as evidence of no effect and (c) the lack of *a priori* sample size calculations. For example, about half the articles that reported statistically non-significant results interpreted them as 'no effect' or 'no relationship', but statistical power calculations were only conducted in two cases. In addition, effect size estimates and associated measures of precision were not presented for many tests, and were infrequently interpreted when they were reported.

The potential consequences of incomplete statistical reporting, particularly low and unknown statistical power or precision, are serious. In applied ecology and, especially, conservation biology it can result in direct, unanticipated, unacceptable environmental damage. We cannot document specific consequences of the misinterpretations and absences detected in our survey, but possibilities include the failure to act when action was warranted, unnecessary expenditure when action was not warranted, and provision of incorrect advice that affected policy and planning decisions. The severity of potential consequences in the field of conservation biology in particular mean that researchers must provide statistical information in a transparent and accessible way, and be highly assiduous in developing professional standards to remedy reporting deficiencies.

An Alternative Focus for the Analysis and Presentation of Ecological Data

In many contexts, interval estimation has considerable advantages over testing zero null hypothesis and reporting results using P-values. The advantages stem from what we feel is the central objective of most ecological studies: to determine if an observed effect (change, impact) is large enough to be important in the context of the ecological system under investigation. Put another way, ecologists are primarily interested in determining if an observed effect is ecologically important, not if it is statistically significant. While P-values provide information about the latter, questions about ecological importance are best answered by estimating the size of effects, and presenting this information with an associated measure of precision. Assuming they are interpreted correctly, there is nothing inherently wrong with presenting P-values along with effect sizes and their associated errors, but P-values alone cannot be used to infer ecological importance. Our view is that, in many situations, P-values should not be reported as they provide no extra information and act as distractions from questions ecologists are attempting to answer.

In this section, we present an example to show how estimates of effects and associated measures of precision can be used to interpret ecological data. Although precision can be calculated in a number of ways, we prefer confidence intervals because we think they are easier to interpret than other measures, and the remainder of this discussion will relate to them. Our example is based on data presented in (Wade 2000) who generated a hypothetical data set to demonstrate the difference between a classical and a Bayesian analysis. Using these data provided an opportunity to contrast the classical approach described by Wade, an approach using confidence intervals, and a Bayesian analysis. We then use a hypothetical example to show how confidence intervals can be used as a planning tool for ecological studies. First, however, we briefly describe what confidence intervals mean and how they should be interpreted.

Meaning and Interpretation of Confidence Intervals

Imagine a survey designed to estimate if the abundance of species X living in recently degraded habitat patches (for example, resulting from timber harvesting or mining) had fallen below a critical level. From extensive past sampling it is known that abundance in surrounding undisturbed habitat has a precisely estimated mean of 20 individuals per hectare, and, based on available information about the reproductive strategy of the species, there is concern that local extinction may occur if abundance in degraded patched falls to about half this level. These considerations guide an *a priori* management decision that remedial action should be taken if abundance falls below 10 individuals per hectare. Initially, a pilot study is established and abundance is measured at five degraded sites. Hypothetical abundance data are 16, 8, 11, 7, and 18 per hectare, and the mean effect size (12.0) and 95% confidence interval (5.98 to 18.02) are shown in Figure 1. Also shown is the P-value function for the confidence interval, a concept we will explain shortly.

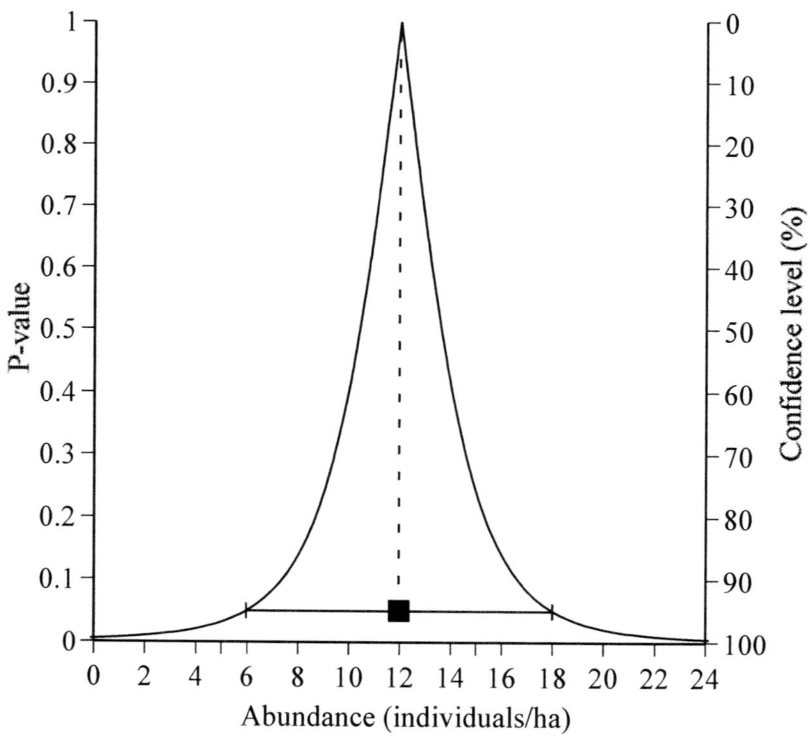

Figure 1. Estimated mean effect (12.0) and 95% confidence interval (5.98 to 18.02) for hypothetical abundance data. The solid line is the P-value function which specifies the consistency of the sample data with each possible abundance value. The vertical dashed line indicates that the data are most consistent with an abundance of 12. Note that P-values (left Y-axis) are directly related to the confidence level (right Y-axis), as is illustrated by the lower and upper limits of the 95% confidence interval cutting the P-value function at 0.05.

Now imagine that this survey was repeated many times, and a 95% confidence interval for the mean calculated on each occasion. Because each sample of data is different, the mean effect and its associated 95% confidence interval will be different on every occasion, but on average, 95% of the intervals will contain the true (unknown) population effect. A common misinterpretation is that we can be 95% sure that an interval from a single sample (eg, 5.98 to 18.02) will contain the population effect, but this is not the case. As with all classical statistical outputs, confidence intervals should be interpreted in the context of notional repeated outcomes.

In many situations, confidence intervals are completely compatible with traditional hypothesis testing procedures - if a 95% confidence interval does not include the value specified by the null hypothesis (usually zero), the null can be rejected at the 5% level (Steidl and Thomas, 2001). In the above example, we know that a test of the null hypothesis that the population effect equals zero will have a P-value less than 0.05 because the confidence interval excludes this null (actually, P = 0.005). In an ecological context, however, the null of interest is 10, and a formal test of this null gives a P-value of 0.41. Although the exact value cannot be determined from the confidence interval alone, we know that the P-value will be large as the null value is near the centre of the interval.

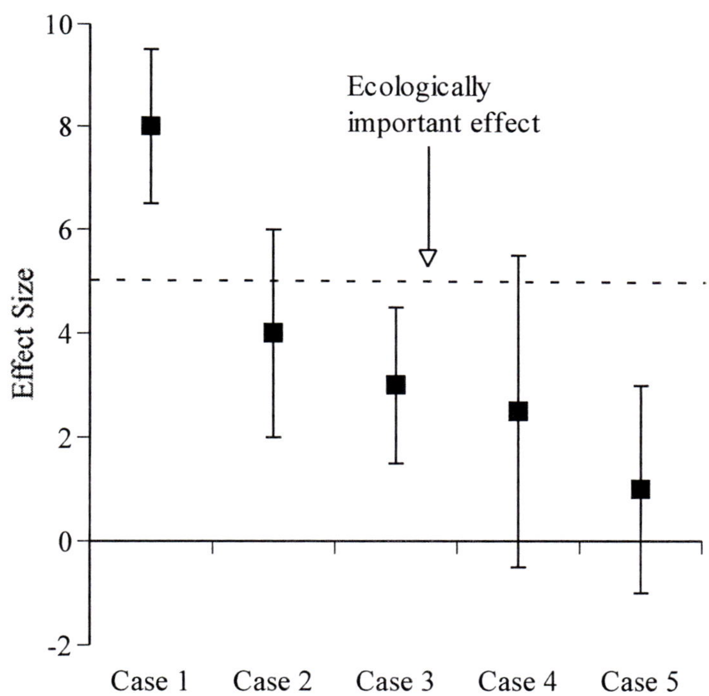

Figure 2. Interpretation of results using confidence intervals. Black squares are estimated effects, error bars are 95% confidence intervals and the dashed line represents an ecologically important effect. In Case 1, the observed effect is both statistically significant and ecologically important. In Case 2, the effect is statistically significant, but the data are insufficient to determine ecological importance. In Case 3, the effect is statistically significant but not ecologically important. In Case 4, the effect is not statistically significant and the data are insufficient to determine ecological importance. In Case 5, the effect is neither statistically significant nor ecologically important. After (Fox 2001) and (Steidl and Thomas 2001).

If researchers can define an ecologically important effect *a priori*, the information provided by confidence intervals can lead to one of five alternative conclusions that differentiate between statistical significance and ecological importance (Figure 2). In many cases, the interpretation of results using confidence intervals will be more complex than this, and will contain a degree of subjectivity (see Example 1 below). Nevertheless, the use of confidence intervals enables researchers to differentiate between statistical significance and ecological importance and thus interpret their data in an ecologically meaningful way.

So what do confidence intervals from a single data sample tell us about the true population effect? In the above example, the confidence interval can be interpreted as representing a range within which the true effect may plausibly lie (Hoenig and Heisey 2001). Another way of putting this is that, although the estimated effect is 12.0, the data are statistically consistent with effects between 5.98 and 18.02 (Goodman and Berlin 1994), although effects closer to the point estimate are most plausible.

This can be shown with that aid of a P-value function ('witches hat' in Figure 1) which assigns a P-value to every possible value of the population effect (which can be thought of as an infinite array of possible nulls).[4] Remember that a P-value derived from testing a zero null hypothesis is defined as the probability of observing the sample data (or data more extreme) assuming the null hypothesis is zero. But, as in the example above, a zero null is infrequently of ecological interest. A P-value function simply displays the P-values related to every possible null value, from zero through to infinity (Rothman 2002). Another way of thinking about this is that a P-value function shows the compatibility of the sample data with each possible null value. Consider the data in Figure 1. The sample data are clearly most consistent with a population effect of 12 ($P = 1.0$), but not at all consistent with population values below 6 or above 18. Values under the peak of the curve (ie, values close to 12) are all reasonably consistent with the observed data while values under the tails of the curve are less consistent with them. The shape of the curve shows the rate at which values within the confidence interval become less consistent with the sample data. If desired, a P-value can be derived for any null value of interest (in the above example, $P=0.41$ for a null value of 10). As we will show in Section 4.2, this can be used to inform confidence interval based inference.

Confidence intervals also provide information about the repeatability of results. Given an original mean and 95% confidence interval, the average probability that a subsequent mean derived from the same population (a replication mean) will fall within the original confidence interval is about 83% (Cumming *et al.* 2004). This is substantially lower than the 95% that might be expected, and arises due to the presence of two sources of variability, one from the original sample and another from the replication sample. Although 83% is the average probability, this value may be higher or substantially lower for a particular sample of data. Figure 3 shows a simulation of 20 independent samples drawn from the same hypothetical population as the pilot data described above. If the mean of a particular sample happens to fall very close to the population value (for example, the third sample mean from the left), and that sample has an interval of typical width, the confidence interval will capture around 95% of replication means. For sample means that happen to fall a little distance away from the population value, however, the confidence interval is likely to capture distinctly fewer. In the

[4] Although we believe that the P-value function is a useful aid to data interpretation, we do not suggest it should be routinely published. Rather, it should be used behind as an interpretive guide. We show it here, however, as we believe it is unfamiliar to most ecologists.

extreme case of a confidence interval that does not capture the population effect (5% of 95% intervals; see the open square cases in Figure 3), less than 50% of future replication effects will be captured.

It is somewhat frustrating that confidence intervals do not provides us with the probability that the interval contains the true effect, a value that would be particularly useful – to achieve this we would have to create intervals using a Bayesian approach. Within the classical statistical framework the best we can do is to say that a 95% confidence interval is very likely but not certain to contain the true effect.

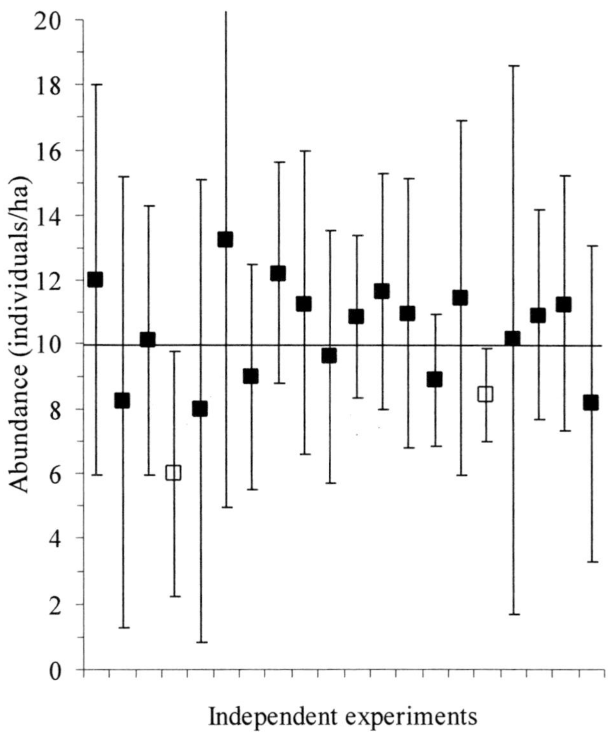

Figure 3. Results of 20 independent samples drawn from the same population as the pilot data. For the purposes of this illustration we assume that population to have mean = 10, standard deviation = 4, and we take samples of size 5. The horizontal line is the population mean, which here is 10 but in practice is never known. Means of successive samples are shown as squares, each with a 95% confidence interval. The first sample is our pilot data. Running a single experiment is equivalent to choosing a single mean and 95% confidence interval randomly from an infinitely large set of means and confidence intervals, just 20 of which are shown here. Note that most but not all confidence intervals capture the population mean; in the long run, 95% of the intervals would capture it. Here, two confidence intervals do not include the population mean, and those sample means are shown as open squares. The figure illustrates that the population mean is more often captured near the centre of a confidence interval than near its extremities. Note that these intervals vary considerably in width because the sample size in our hypothetical example is small

Example 1: Using Confidence Intervals to Interpret Ecological Data

(Wade 2000) generated data sets for two hypothetical animal populations (population 1 and population 2; Figure 4) to explore the differences between a classical hypothesis testing approach and a Bayesian analysis. Wade presented graphs of the data including the line of best fit, the regression equations (Figure 4 caption) and summarised the results of the hypothesis tests (null hypothesis of zero slope) using P-values (P = 0.048 for population 1 and P = 0.053 for population 2). In addition to Wade's data presentation and analysis we have drawn 95% confidence intervals around the line of best fit (dashed lines in Figure 4) and calculated 95% confidence intervals around the slopes; the slope (95% CI) for population 1 is -0.36% (-0.004 to -0.72%) and for population 2 is -10.0% (0.19 to -20%). In addition, we have calculated a P-value function for these intervals to aid interpretation (Figure 5).

In a conservation context, data like those in Wade's hypothetical example are collected to detect if a population is declining at an ecologically important rate. For the purposes of his example, (Wade 2000) assumed that a decline of 5% per year would be ecologically important, and we will adopt the same value here.

In his exposition of the classical hypothesis testing approach, (Wade 2000) assumed that management intervention to halt a decline would only occur if the gradient of the regression line was statistically less than zero. Operating within this framework, population 1 is assumed to be declining (P = 0.048) and deserving of management action, while population 2 is not (P = 0.053). As Wade has said, this conclusion is irreconcilable with the data. Simply looking at the graphs (Figure 4) clearly shows that population 2 may well be declining at a much greater rate than population 1.

This example raises two further issues regarding the dichotomous decision-making approach described above. First, it demonstrates the nonsensical nature of an arbitrary cut-off point (in this case a P-value of 0.05) for decision-making purposes. Because both P-values are so close to 0.05, a small change in either data set could easily result in a different decision, even though the overall trends would be substantially the same. We would not use P-values to interpret these data, but if presented with nothing else we would conclude that the two data sets provide approximately equal statistical evidence for a population decline. Second, the data from population 1 are precisely estimated and indicate a small decline while the data from population 2 provide a much less precise estimate and indicate a decline that may be small or very large. The calculation of the P-value requires both these pieces of information (the precision and the effect size) and so their independent influence is hidden. As this example shows, the combination of small effect and high precision in population 1 and large effect and low precision in population 2 produce approximately equal P-values, even though the data from the two populations are of a very different form. The estimated effect and the precision of this estimate are important in their own right and should both be reported.

(Wade 2000) dismissed the hypothesis testing approach described above as inappropriate and advocated a Bayesian solution to the problem. Using an uninformative prior distribution, Wade's Bayesian analysis indicated that the probability of a decline greater than 5% per year is 0 for population 1 and 0.86 for population 2. In addition, the probability that the decline is between 0 and 5% is 0.98 for population 1 and 0.12 for population 2. These data suggest that population 1 is very likely to be declining, but at a rate less than 5% per year while population 2 is probably declining at a rate of greater than 5% per year. We agree that this is a sensible way to analyse the data and the conclusion would be an appropriate guide for management action.

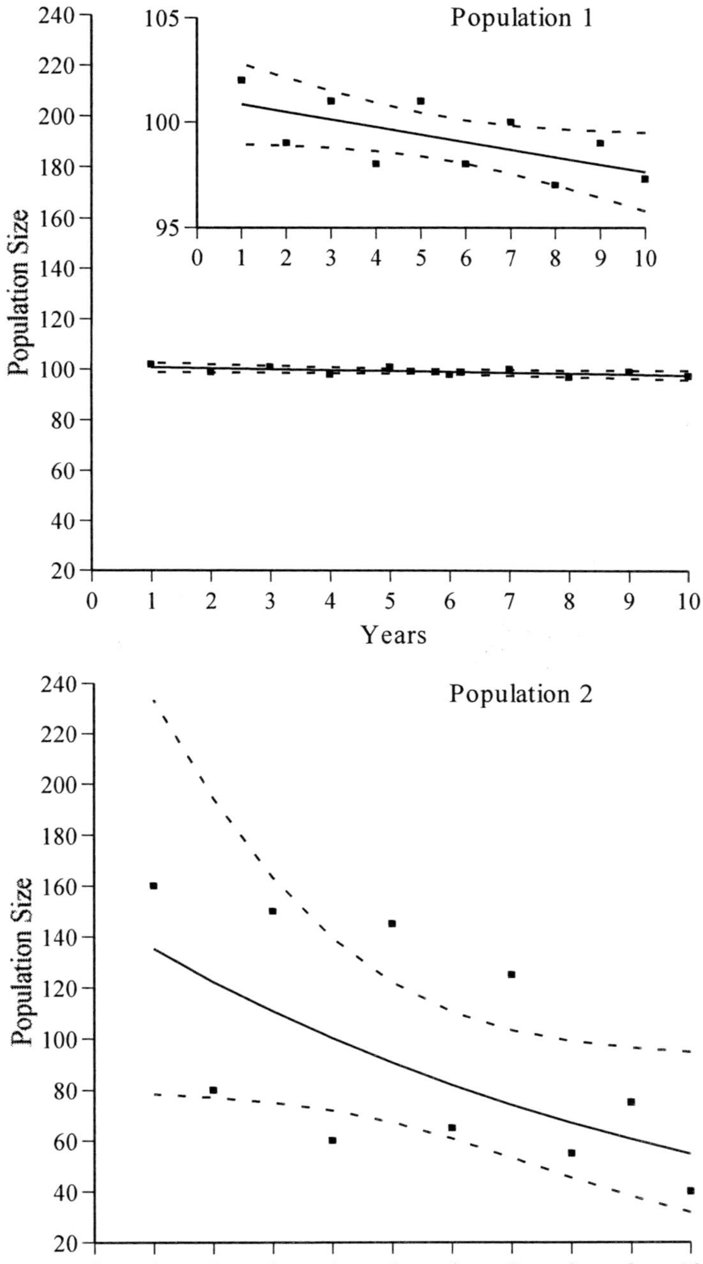

Figure 4. Population 1 and population 2 from (Wade 2000). The solid line is the line of best fit and the dashed lines are 95% confidence intervals (data have been back transformed for easier interpretation). The main graphs show the data at the same scale for the purpose of comparison, but population 1 data are enlarged in the inset. Population 1: logey = 4.62 – 0.0036x; 95% CI for the slope = -0.00004 to –0.007; P = 0.048; Population 2: logey = 5.01 – 0.10x; 95% CI for the slope = 0.002 to –0.20; P = 0.053

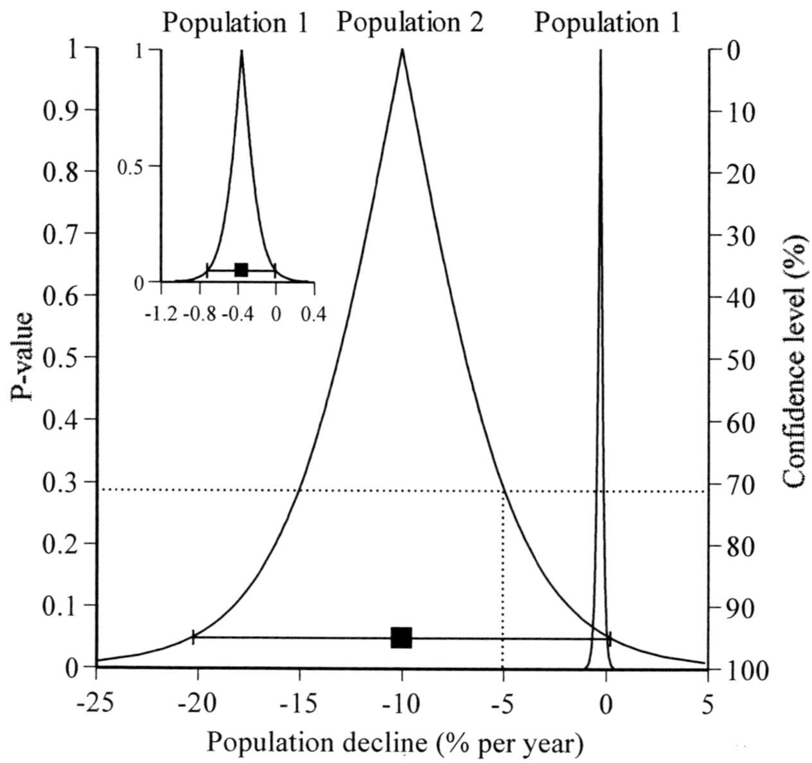

Figure 5. P-value functions (solid lines), estimated population declines (black squares) and associated 95% confidence intervals (error bars) for populations 1 and 2. The inset shows data for population 1 on a readable scale. For population 2, the dotted lines show the P-value and confidence level corresponding to a decline of 5% per year. The P-value is 0.29 and the confidence level is 71%

Interval estimation, however, can also be used to interpret these data in a rational way. The statistical technique is classical, but the focus is on effect size estimates (the regression slopes) and the associated confidence intervals.

In percentage terms, population 1 is estimated to be declining at 0.36% per year, but the 95% confidence interval indicates that the data are statistically compatible with declines between 0.004% and 0.72% per year. Because the confidence interval does not include a decline of 5%, we conclude that the data are not statistically compatible with a decline of this magnitude (Case 3 in Figure 2). This is clearly demonstrated by the P-value function (Figure 5) which shows that the data are most consistent with a small decline, but completely inconsistent with a decline greater than about 1% per year (P=0 for declines greater than this value). In contrast, population 2 is estimated to be declining at 10% per year, but the data are statistically compatible with a slight population increase (0.19%) or a decline of up to 20% per year. Population 2 is almost certainly declining, but because the data are variable, we cannot predict the true rate of decline with precision. We feel there are three possible ways to deal with this kind of situation.

a. Acknowledge that the population is almost certainly declining, and may well be declining at >5% per year, but conclude that more data are needed before a decision can be made regarding management intervention. We might, for example, specify that we require enough data for the upper bound of the 95% confidence interval (the bound closer to zero) to exclude a 5% per year decline (Case 1 in Figure 2) before recommending management action. This approach is conservative and weighted towards a conclusion that a critical decline is not occurring.

b. Accept that the effect size estimate is imprecise, but recommend management intervention because an ecologically important decline may well be occurring. This conclusion is based on the fact that (a) the best estimate of the decline is 10% per year, and (b) the ecologically important decline (5% per year) is in the right hand tail of the P-value function (Figure 5) making larger (more negative) declines much more probable than smaller declines. As illustrated in Figure 5, the P-value associated with a decline of 5% per year is 0.29, and increases rapidly for larger values. We interpret this to mean that while declines less rapid than 5% per year are plausible values for the true (population) decline, declines more rapid than 5% per year (declines in the range of 8 to 12%, for example) are much more consistent with the sample data.

c. Accept a higher level of error and use a lower confidence level (eg, 90% or 80%). From the P-value function it can be seen that a 71% confidence interval would just exclude the ecologically important effect. There is strong evidence that in many conservation scenarios it is more sensible to weight the decision process in favour of effect detection (Fairweather 1991; Peterman 1990; Taylor and Gerrodette 1993), so using a lower confidence level may well be justified. Having said this, we believe that the confidence level should be kept as high as possible, and, if it is lowered, this should be done during the planning phase of a study. We provide some additional comments in Section 4.3.

We feel that option (b) above is the best approach. The confidence interval clearly shows that the point estimate lacks precision and the sample data are statistically consistent with a wide range of possible population declines. The P-value function, however, demonstrates the consistency of various population declines with the sample data, and indicates that declines between zero and 5% per year are much less consistent with the data than a range of declines greater (ie, more negative) than 5% per year. Thus our conclusion matches that of Wade's Bayesian analysis; population 2 is probably declining at a rate greater than 5% per year, and management intervention should occur.

It is important to point out that the inferential process we have just described contains two decisions based on judgement.[5] The first is the use of a 95% confidence interval (as opposed to a 90% or 80% interval), and the second is the way we interpreted the P-value function as evidence for a >5% decline. Reasoned subjectivity is a necessary part of all data analytic processes whether we like it or not (Stewart-Oaten 1995). The Bayesian approach described by (Wade 2000) also contained judgements. Wade, for example, decided to use an uninformative prior distribution, but it would have been quite acceptable to use an informative prior, a decision that would have altered the output of the analysis. In addition, Wade describes the probability of population 2 declining at >5% (0.86) as 'fairly high', and

[5] In the context of a real study, many judgements concerning the experimental design, data collection protocols, measurement variables, etc, are made prior to analysis. All of these decisions have the capacity to influence the results.

indicates that the Bayes factor comparing the hypotheses 'slope = zero' to 'slope = -5%' provides 'positive' evidence of a 5% decline. Wade concludes (and we agree) that, on the basis of the available evidence, population 2 is probably declining at a rate of at least 5% per year, but this is a reasoned judgement and not an indisputable fact.

In reality, ecological data are highly variable, and, combined with the small sample sizes associated with many ecological research projects, effect size estimates are frequently surrounded by large error margins. Using confidence intervals to interpret ecological data enables this error to be quantified and displayed in a form that is readily appreciated, and facilitates its transparent incorporation into the decision-making process.

Example 2: Using Confidence Intervals to Plan Ecological Studies

It is widely acknowledged that power analysis is a useful tool in the planning phase of ecological studies (Di Stefano 2001, 2003; Downes *et al.* 2002; Fairweather 1991; Foster 2001; Gerrodette 1993; Keough and Mapstone 1997; Quinn and Keough 2002; Steidl and Thomas 2001), but the utility of confidence intervals for study planning has rarely been discussed in the ecological literature (but see (Steidl and Thomas 2001)). Nevertheless, confidence intervals have been considered as a planning tool in other disciplines (Goodman and Berlin 1994; Smithson 2003) and we feel they are just as applicable to ecology.

The use of confidence intervals for study planning has been discussed within both hypothesis testing (Daly 2000; Goodman and Berlin 1994; Smithson 2003; Steidl and Thomas 2001) and interval estimation (Algina *et al.* 2002; Algina and Olejnik 2000) frameworks. The former involves determining sample size using statistical power analysis and then substituting this value into a confidence interval equation to examine the corresponding expected confidence interval width. We refer to this as the power approach. The alternative is simply to use a confidence interval equation to calculate the sample size needed to generate an interval of specified width. We refer to this as the precision approach. As has been noted elsewhere (Algina *et al.* 2002; Algina and Olejnik 2000; Smithson 2003) these methods often produce quite different answers, because the sample size necessary to test a hypothesis with adequate power and to estimate an effect with adequate precision may well be different.

There is in addition a more fundamental distinction between the two approaches. Power analysis is inextricably linked to statistical significance testing and requires the specification of point null and alternative hypotheses. The difference between these hypotheses is the 'important effect size' integral to all power calculations, and outputs of this process can only be interpreted in relation to the effect size value used in the calculation. In contrast, the precision approach requires a specification of the desired precision (defined as w, the confidence interval half width, also known as the *margin of error*) with which to estimate the parameter of interest. Consequently we can use this method to calculate the sample size required to estimate a parameter with a specified degree of precision regardless of its true population value. If we were using the power approach, a value for the true population parameter would need to be specified.

Although the simplicity of the precision approach seems appealing, it has been criticised for resulting in sample sizes that may be considered too small, as the probability that the interval width actually observed in a future study will be greater (or less) than the planned width is approximately 50% (Daly 2000; Goodman and Berlin 1994). In many research scenarios, only having a 50% chance of obtaining one's specified level of precision may be unacceptable. Due to this problem, most of the literature on this topic suggests using power analysis for at least some components of the study planning process, even if confidence intervals will be used to interpret the results. However, we feel the use of statistical power analysis to determine sample size is inconsistent with interval estimation philosophy, and a modification of the precision approach is required.

Here we suggest a novel precision-based method for determining sample size that provides a solution to the problem noted above. For the purpose of comparison, we demonstrate the process using both statistical power analysis and our novel precision method with reference to the hypothetical example outlined in Section 4.1.

The power approach. Let us return to the example outlined in Section 4.1 (Figure 1). Recall that the objective was to determine if abundance at degraded sites had fallen below a critical level which, on the basis past data and informed opinion, had been set at 10 individuals per hectare. As the data from the pilot study were inconclusive, we may wish to know how many degraded sites need to be surveyed to be confident that abundance has or has not fallen below the critical level.

To perform the power calculation we need to specify an important effect size, values for α and power, and use the standard deviation of the pilot data as an estimate of the population standard deviation. In this case, the important effect size is the smallest difference between the future sample mean and our critical value that we think it is important to detect. We might choose this value to be 2 as we are prepared to accept that abundance estimates between 8 and 12 (within 20% of the critical value) are functionally equivalent. Consequently, the null hypothesis is that abundance at degraded sides is equal to 10, while the alternative is that abundance is 2 units away from 10, in either direction. We are interested in this two tailed alternative as, for management purposes, it is important to determine whether abundance is less than or greater than the critical level.

Rather than deciding on specific values of α and β we follow the advice of (Keough and Mapstone 1997; Mapstone 1995) and others and determine a relevant α:β ratio by considering the relative cost of Type I and Type II errors. From a management perspective, making a Type I error would mean investing resources to increase abundance in degraded patches when such an investment was not necessary, while making a Type II error would mean there would be no management action even though it was warranted, resulting in the possible local extinction of the species under consideration. For the purpose of this example let us assume that after thorough discussion with all relevant stakeholders, it is decided that Type I and Type II errors are equally costly, thus defining the α:β ratio as 1:1.

Because specific levels of α and power are not defined, we start with a high value for power. Entering power = 0.95 and α = 0.05 into a suitable power calculator (Lenth 2000), the number of degraded sites that need to be surveyed to detect an effect size of 2 is 78. Inserting n = 78 into the appropriate confidence interval formula gives an expected 95% confidence interval half width of 1.09, although the probability of observing a value of w (the confidence interval half width) no greater than this in future data is only about 50%. If there were not 78 degraded sites or the money was not available for such extensive sampling effort, we would reduce power and increase α so that the 1:1 α:β ratio is maintained. For example, when power = 0.90 and α = 0.10, 52 degraded sites are required, and the corresponding 95%

confidence interval half width is 1.35, and when power = 0.80 and $\alpha = 0.20$, we need 27 sites and the interval half width is 1.92. Until now, this kind of process has been the best way to explore the relationship between sample size, statistical power and confidence interval width.

The Precision Approach

If the half width of a confidence interval is w, $w = t_{(df),C} \times SE$, where t is the two tailed critical value from a t-distribution with df degrees of freedom, C is the confidence level and SE is the standard error. Further, we differentiate between w_{exp} and w_{obs}, the expected (as calculated pre-study) and observed (post-study) values of w respectively. We use the fact that w is calculated from the sample standard deviation whose sampling distribution is related to the chi-square distribution. We can therefore calculate the one tailed w_{upper} value for any chosen probability PP so that w_{obs} will be less than w_{upper}, for a specified C and sample size. We use PP for the *precision probability*, the probability that our observed precision (w_{obs}), which varies from sample to sample, will be no greater than w_{upper}. PP can also be thought of as the chance that the observed confidence interval will be at least as precise as planned. This process provides a calculation based on the trade-off between n and w_{upper}, for our chosen PP. No null or alternative hypotheses need be specified, but if we do wish to consider precision in relation to some chosen effect size, we could choose w_{upper} equal to that effect, then calculate the corresponding sample size.

For our calculations, we might specify our desired precision, w, to be 2, for similar reasons as outlined in the power example. We then might want to know the sample size required to have a PP% chance that the observed precision, w_{obs}, will be less than 2 when the confidence level, C, is set at 95%. Using the precision approach, we find that 25 degraded sites are required to have a PP = 50% chance of obtaining w_{obs} less than 2 in a future sample. For PP = 80% and 95%, the required sample size is 30 and 35 respectively. If we wished, we could specify different levels of precision and investigate the affect on sample size. For example, if $w = 1$ and C = 95%, the sample sizes required to achieve a PP of 50%, 80% and 95% are 93, 104 and 114 respectively. Using the standard deviation estimate from the pilot data, Figure 6 shows the relationship between sample size and PP for w = 1, 2 and 3. C is 95% in all cases.

In summary, our novel precision approach to study planning enables researchers to vary the precision probability, PP, and thus examine the relationship between this value, sample size and a specified degree of precision. We believe this method for sample size determination is philosophically consistent with the interval estimation approach advocated in this chapter, and by enabling researchers to vary the precision probability we provide a solution to the problem of inadequate sample size that has been the basis of criticism in the past. If they wish, researchers can also vary the confidence level, C, and observe how this affects the outcome. If possible, however, C should remain high, as the lower C becomes the less confident we are that values within the confidence interval contain the population parameter. Although we recognise that smaller values of C may be relevant in some situations, we encourage researchers to keep the confidence level as large as possible.

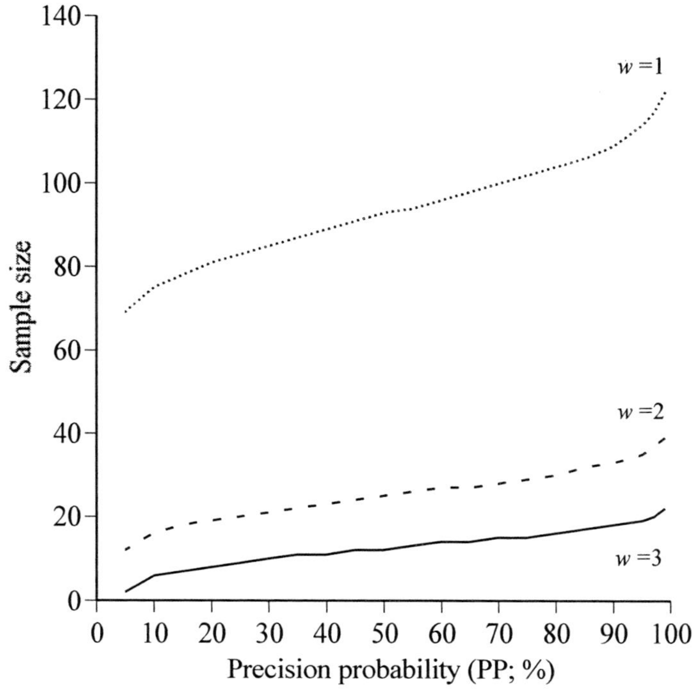

Figure 6. The relationship between sample size and the precision probability (PP; the chance that the confidence interval half width observed in a future sample will be less than a specified value, w) for three different values of w. In this case, the confidence level (C) = 95% and standard deviation = 4.85.

Using Confidence Intervals Wisely: Some Identified Misconceptions and Guidelines for Interpretation

In the previous section we described some advantages associated with a confidence interval approach to interpreting ecological data. However, confidence intervals are not a panacea to all statistical reporting problems; they too are misinterpreted, and the extent to which they are able to supplement or replace P-values is not clear. Although research in cognitive psychology has explored many of the misconceptions associated with P-values and tests of zero null hypothesis (Oakes 1986; Tversky and Kahneman 1971), corresponding work on confidence intervals is relatively undeveloped. Claims regarding the benefits of confidence intervals include the relative ease with which they are understood and taught, their capacity to facilitate better communication of research results and their infrequent misinterpretation (Schmidt 1996; Schmidt and Hunter 1997). Others admit that confidence intervals may be misinterpreted, but that the consequences of such misinterpretations are likely to be less serious than corresponding errors associated with P-values (Hoenig and Heisey 2001). Nevertheless, these statements contain no supporting evidence and the empirical questions they raise are largely unanswered. It is clear that additional research in this area is required.

Avoiding Established Misconceptions: 'Statistically Non-Significant Equals No Effect'

One of the misconceptions associated with P-values (Section 3) is that statistically non-significant results (usually a P-value >0.05) are often equated with the absence of an ecologically important effect. To test the hypothesis that the interpretation of results using confidence intervals reduces the occurrence of this error, we presented 79 final year or postgraduate ecology students with two similar research scenarios that reported results using P-values and confidence intervals respectively. The results of both scenarios were statistically non-significant, but there was a clear specification that statistical power was low and that the observed effect sizes were non trivial. Using a five point Likert scale, students were asked to interpret the data as evidence for or against the null hypothesis of no effect (Fidler *et al.*, unpublished data).

When statistically non-significant results were presented using P-values, almost 40% (31 of 79) of students believed that the data provided positive evidence for the null hypothesis of no effect (the wrong answer), despite the presence of statistical power information and additional guidance about the ecological importance of observed effects. However, the overwhelming majority of these students (87%, 27 of 31) changed their mind when they viewed similar results presented using confidence intervals. Interestingly, of the 60% or so that interpreted results correctly when presented with P-values, 17% (8 of 48) misinterpreted the results when they were expressed using confidence intervals. Thus although confidence intervals usually facilitated the correct interpretation of results, they occasionally resulted in the reverse effect.

To test whether this reverse effect was an artefact of different scenario content, the study was repeated so that each student received the same scenario presented once with P-values and once using confidence intervals. This time we tested 55 second year ecology students, asking the same question. The results of this study were almost identical to the first except that the reverse effect was reduced by about half. An additional result was the apparent presence of a learning effect. Students who saw the confidence intervals first gave the correct answer on the P-value presentation more often than students who saw P-values first, but there was no corresponding beneficial transfer of seeing the P-values before confidence intervals.

Relative to research scenarios where results were presented using P-values, the students we tested were less likely to interpret statistically non-significant results as evidence for 'no effect' when confidence intervals were used. Based on these preliminary data, confidence intervals appear to assist the precautionary interpretation of low powered, statistically non-significant results.

Creating New Misconceptions?

In other surveys we have found confidence intervals susceptible to unique misconceptions not commonly associated with P-values. In two separate surveys of undergraduate students in psychology and ecology (total n=357), we found that many failed to recognise confidence intervals as inferential (Fidler and Finch, unpublished data). In a variety of contexts and question types most students interpreted confidence intervals as relating to the data sample from which they were derived, but not to the relevant population parameter. In students' own definitions, they described the confidence interval as a range or a truncated range of sample observations or measures, rather than as a plausible set of values within which the population parameter is likely to lie.

It is not surprising that undergraduate students have a developing understanding of statistics and thus misinterpret confidence intervals to some degree. Recent research, however, shows that misconceptions about confidence intervals, along with standard error bars, extends to experienced researchers in a number of disciplines, including medicine where confidence intervals are routinely used. Three of these interpretive errors are outlined below.

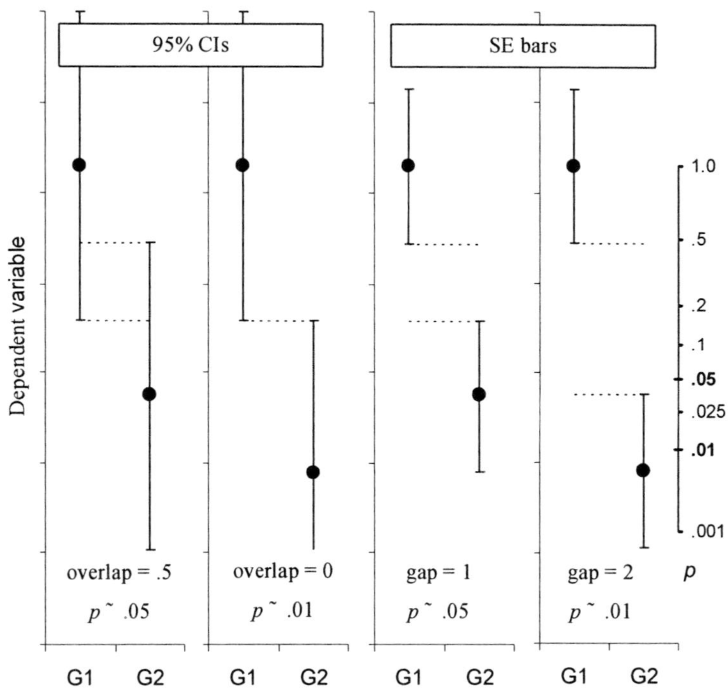

Figure 7. Guidelines for interpreting the difference between the means of two independent groups G1 and G2. The left two panels show 95% confidence intervals, with overlap 0.5 and 0, and the corresponding approximate P-values, 0.05 and 0.01 respectively. (Overlap of 0.5 means overlap equal to half the length of one arm of either confidence interval.) The right two panels show standard error bars, with gap 1 and 2 standard errors, and corresponding approximate P-values, 0.05 and 0.01 respectively (one standard error is one arm of the standard error bar). The vertical scale at right shows the P-value as a function of the G2 mean, all other aspects remaining the same. This scale, which applies to all four panels, shows that the P-values stated are a little conservative - the precise values in this case are a little lower than 0.05 and 0.01

The Overlap Misconception

Many researchers falsely believe that for two independent group means to be statistically significantly different, the 95% confidence intervals around those means must not overlap (Schenker and Gentleman 2001). In fact, 95% confidence intervals can overlap by roughly a quarter and the mean difference still be significant at the 5% level (Figure 7). In an internet-based study, (Belia *et al.* submitted) tested the extent of the overlap misconception by inviting authors published in leading psychology, medicine and behavioural neuroscience journals to adjust a figure until they judged two means, each with 95% confidence intervals, to be just statistically significantly different (P = 0.05). Only 17% (24 of 140) of respondents adjusted the figure so that corresponding P-value was close to 0.05. Most respondents were too conservative, setting the confidence intervals to just touch or not overlap at all. The mean response positioned the confidence intervals at the equivalent of P = 0.009.

Drawing from the same population of researchers, (Belia *et al.* submitted) conducted the same experiment when the means were surrounded by standard error bars (roughly half the width of 95% confidence intervals). In this case, 25% (44 of 179) of respondents adjusted the figure so that corresponding P-value was close to 0.05. As with the confidence interval based exercise, most researchers set the means so that arms of the standard error bars almost touched, which in this case led to answers that were too lax; the mean response corresponded to P =0.109. In fact, standard error bars must be separated by about one standard error (ie, one arm of a standard error bar) for the corresponding P-value to be about 0.05 (Figure 7). Further discussion concerning the interpretation of error bar overlap can be found in (Cumming and Finch submitted; Saville 2003; Wolfe and Hanley 2002).

Error Bar Interpretation and Experimental Design

Figure 8 is an example of a graph displaying typical repeated measures data, including control and treatment means, with associated error bars, at a number of different times. Unknown to many researchers is that error bars around individual means are useful only for interpreting data from the same time interval. Only error bars around the *difference between times* allow us to look directly at the effect of time, which is usually of interest in studies containing repeated temporal measurements. With reference to Figure 8, error bars can be used to assess the difference between Control and Treatment means at Time 1, (for example, by using the overlap guidelines in Figure 7), but the error bars are irrelevant, even misleading, for comparing, for example, Treatment means at Time 1 and Time 2.

Again drawing from a population of published authors in psychology, medicine and behavioural neuroscience, (Belia *et al.* submitted) tested the extent to which this issue was understood. Only 11% (18 of 159) of respondents realised that bars around individual means could not used to determine statistical significance in repeated measures cases. It may well be that current graphical conventions promote mistakes of this kind, as adjacent error bars like those displayed in Figure 8 invite invalid comparison between correlated data, and traditional figure design does not make salient whether an independent variable is a repeated measure or not.

The Confidence Level Misconception

This is the mistaken belief that a 95% confidence interval will, on average, capture 95% of effects from future samples (replication effects). As we mentioned in Section 4.1, however, a 95% confidence interval only captures on average about 83% of replication effects. In a second internet-based study, (Cumming *et al.* 2004) contacted researchers and asked them to estimate where 10 replication means might plausibly fall given an observed mean and its 95% confidence interval. Almost 80% (105 of 134) of respondents placed 9 or 10 of the replication means within the confidence interval. In open-ended comments many made clear that they had attempted to match the number of future means within the intervals to the confidence level (ie, 95%). Corresponding results were found in a group asked to perform the same task with standard error bars. Most respondents placed 6 or 7 of the replication means within the standard error limits, again matching the number of means to the confidence interval (68% for standard errors). These results indicate that researchers often have a poor understanding of the information error bars provide about where replication means are likely to fall.

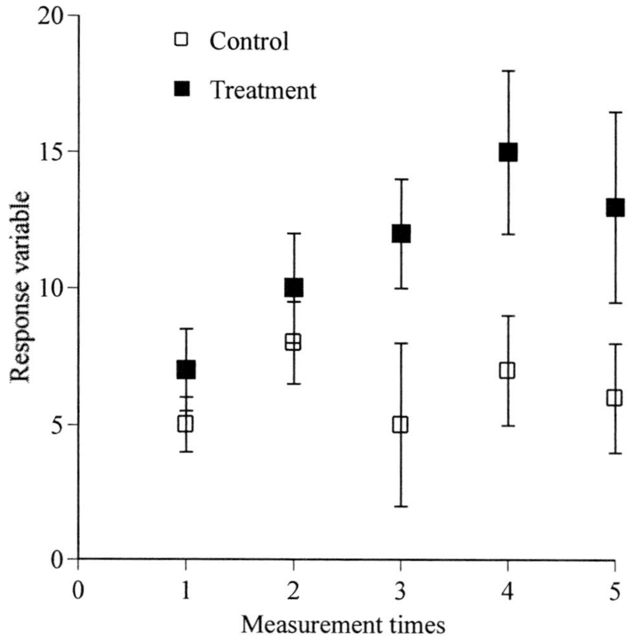

Figure 8. Hypothetical data representing a design that includes a repeated measure. The error bars (whether they are confidence intervals or standard errors) can only be used to draw inferences about control and treatment means at the same measurement time and cannot be used to infer temporal differences or trends. The guidelines of Figure 7 may be used to make inferences about independent groups, for example Control vs Treatment means at Time 1, but may not be used to compare a repeated measure, such as Treatment mean at Time 1 vs Treatment mean at Time 2.

Guidelines for Interpreting Confidence Intervals

Studies conducted to date suggests that a number of misconceptions and interpretive problems may occur when confidence are used to interpret research results. Although these studies have investigated responses of scientists in psychology, medicine and related fields, we expect the same interpretive errors would be made by ecologists. Despite this, we encourage the use of confidence intervals as we believe they are superior to tests of zero null hypotheses and their associated P-values. To support the increased use of confidence intervals we present a number of guidelines designed to minimise interpretive errors. Further guidance for interpreting confidence intervals can be found in (Cumming and Finch submitted).

- Consider each value within a 95% confidence interval as a plausible estimate of the population parameter, but, although it is probable, there is no guarantee that the population parameter lies within the interval.

- Note that the population parameter is more likely to be near the centre of a confidence interval, near the point estimate, and less likely to be near the outer limits.
- Consider whether a confidence interval contains an ecologically important effect. If so, look at where it is positioned. As mentioned above, effects closer to the centre of the interval are more plausible.
- For a comparison of two independent means, $P \approx 0.05$ when the arms of the 95% confidence intervals overlap by about half the length of one arm, that is, by about one quarter of the total confidence interval width. $P \approx 0.01$ when the overlap is zero or there is a gap between the two intervals. (This applies when both sample sizes are at least 10, and the two interval widths are not substantially different; see Figure 7).
- In studies that include a repeated measure, such as those in which measurements are taken at several times, interval estimates (whether confidence intervals or SE bars) on means at individual times may not be used to guide inferences that involve the repeated measure. See Figure 8.
- Note the width of the interval and what this tells you about the precision of the study. A wide interval indicates that the population parameter has been estimated imprecisely. In general, studies with wide confidence intervals have low statistical power to detect effect sizes likely to be of interest.

CONCLUSIONS

Based on surveys of the literature it is clear that tests of zero null hypotheses and the use of P-values to make dichotomous decisions about the significance or otherwise of results is widespread in ecology. However, we feel there is good reason for ecologists to consider alternative approaches. The approach outlined here, interval estimation, involves estimating effects and presenting them along with their associated confidence intervals. When confidence intervals are reported, time should be taken to interpret and discuss the information they convey. Wherever possible, ecologists should determine ecologically relevant effect sizes prior to data collection and use these values to aid sample size determination. We believe that interval estimation is highly compatible with the analytical objectives of most ecological studies and thus facilitates a transfer of accurate information to both investigators and applied users.

Proponents of interval estimation often assume that this technique is free of the problems associated with hypothesis testing. This assumption, however, has not been rigorously tested. The preliminary data presented in this chapter indicate that confidence convey more meaning than P-values, but a number of misconceptions still occur. In many instances, the cognitive processes associated with the interpretation of confidence intervals are unclear. In addition, there is some debate regarding the best way to use confidence intervals for study planning, and we contribute to this by suggesting a novel strategy based on precision. Clearly, a greater degree of consideration and research into these aspects of interval estimation is required.

We believe that journal editors have a major role to play in encouraging the use of interval estimation and other sensible data analysis strategies. Editorial policy outlining the benefits of *a priori* sample size calculations, the specification of ecologically important effect sizes and the use of confidence intervals to report the error associated with point estimates would send a strong message to potential authors that these practices are preferred. While some ecology journals have such guidelines (*Ecology*, and the other Ecological Society of America publications, for example), most do not. As (Ellison 2004) has shown, however, the presence of guidelines do not solve the problem of misinterpreting and misreporting P-values, and comments on manuscripts by editors and reviewers should alert authors to these problems. In one sense, editors are the gatekeepers of ecological data, and thus have tremendous potential to improve statistical reporting practices across the discipline.

ACKNOWLEDGMENTS

Russ Lenth, Graham Hepworth, Paul Wade, Lauren Bennett, Mark Burgman, Kylie McKenzie, This work was supported by the Australian Research Council.

REFERENCES

Algina J., Moulder B. C. and Moser B. K. (2002) Sample size requirements for accurate estimation of squared semi-partial correlation coefficients. *Multivariate Behavioral Research* 37: 37-57.

Algina J. and Olejnik S. (2000) Determining sample size for accurate estimation of the squared multiple correlation coefficient. *Multivariate Behavioral Research* 35: 119-136.

Anderson D. R., Burnham K. P. and Thompson W. L. (2000) Null hypothesis testing: Problems, prevalence, and an alternative. *Journal of Wildlife Management* 64: 912-923.

Anderson D. R., Link W. A., Johnson D. H. and Burnham K. P. (2001) Suggestions for presenting the results of data analyses. *Journal of Wildlife Management* 65: 373-378.

Barrick K. A. (2003) Comparison of the nutrient ecology of coastal *Banksia grandis* elfinwood (windswept shrub-like form) and low trees, Cape Leeuwin-Naturaliste National Park, Western Australia. *Austral Ecology* 28: 252-262.

Bayes T. (1763) An essay towards solving a problem in the doctrine of chances. *Philosophical Transactions of the Royal Society* 53: 370-418.

Belia S., Fidler F., Williams J. and Cumming G. (submitted) Researchers misunderstand confidence intervals and standard error bars. *Psychological Methods*.

Berger J. O. and Sellke T. (1987) Testing a point null hypothesis - the irreconcilability of P-values and evidence. *Journal of the American Statistical Association* 82: 112-122.

Burnham K. P. and Anderson D. R. (1998) *Model Selection and Inference: a Practical Information-Theoretic Approach*. Springer, New York.

Burnham K. P. and Anderson D. R. (2001) Kullback-Leibler information as a basis for strong inference in ecological studies. *Wildlife Research* 28: 111-119.

Chatfield C. (1985) The initial examination of data. *Journal of the Royal Statistical Society, Series A* 148: 214-253.

Cohen J. (1994) The earth is round (p<.05). *American Psychologist* 49: 997-1003.

Cox D. R. (1977) The role of significance tests. *Scandinavian Journal of Statistics* 4: 49-70.

Cumming G. and Finch S. (submitted) Inference by eye: Confidence intervals, and how to read pictures of data. *American Psychologist*.

Cumming G., Williams J. and Fidler F. (2004) Replication, and researchers' understanding of confidence intervals and standard error bars. *Understanding Statistics (in press)*.

Daly L. E. (2000) Confidence intervals and sample sizes. In: *Statistics with Confidence* (eds. D. G. Altman, D. Machin, T. Bryant, N. and M. J. Gardner) pp. 139-152. BMJ Books, Bristol.

Di Stefano J. (2001) Power analysis and sustainable forest management. *Forest Ecology and Management* 154: 141-153.

Di Stefano J. (2003) How much power is enough? Against the development of an arbitrary convention for statistical power calculations. *Functional Ecology* 17: 707-709.

Di Stefano J. (2004) A confidence interval approach to data analysis. *Forest Ecology and Management* 187: 173-183.

Downes B. J., Barmuta L. A., Fairweather P. G., Faith D. P., Keough M. J., Lake P. S., Mapstone B. D. and Quinn G. P. (2002) *Monitoring Ecological Impacts: Concepts and Practice in Flowing Waters*. Cambridge University Press, Cambridge.

Ellison A. M. (1996) An introduction to Bayesian inference for ecological research and environmental decision-making. *Ecological Applications* 6: 1036-1046.

Ellison A. M. (2004) Bayesian inference in ecology. *Ecology Letters* 7: 509-520.

Fairweather P. G. (1991) Statistical power and design requirements for environmental monitoring. *Australian Journal of Marine and Freshwater Research* 42: 555-567.

Fidler F., Cumming G., Burgman M. and Thomason N. (in press) Statistical reform in medicine, psychology and ecology. *Journal of Socio-economics*.

Fidler F., Thomason N., Cumming G., Finch S. and Leeman J. (2004) Editors can lead researchers to confidence intervals, but can't make them think - Statistical reform lessons from medicine. *Psychological Science* 15: 119-126.

Fisher R. A. (1925) *Statistical Methods for Research Workers*. Oliver and Boyd, Edinburgh.

Fisher R. A. (1935) *The Design of Experiments*. Oliver and Boyd, Edinburgh.

Fisher R. A. (1955) Statistical methods and scientific induction. *Journal of the Royal Statistical Society Series B* 17: 69-77.

Fisher R. A. (1956) *Statistical Methods and Scientific Inference*. Oliver and Boyd, Edinburgh.

Foster J. R. (2001) Statistical power in forest monitoring. *Forest Ecology and Management* 151: 211-222.

Fox D. R. (2001) Environmental power analysis - a new perspective. *Environmetrics* 12: 437-449.

Gerard P. D., Smith D. R. and Weerakkody G. (1998) Limits of retrospective power analysis. *Journal of Wildlife Management* 62: 801-807.

Gerrodette T. (1993) Trends: software for a power analysis of linear regression. *Wildlife Society Bulletin* 21: 515-516.

Gigerenzer G. (1993) The superego, the ego, and the id in statistical reasoning. In: *A Handbook for Data Analysis in the Behavioral Sciences: Methodological Issues* (eds. G. Keren and C. Lewis) pp. 311-339. Lawrence Erlbaum, Hillsdale.

Gigerenzer G., Swijtink Z., Porter T., Daston L., Beatty J. and Kruger L. (1989) *The empire of chance. How probability changed science and everyday life*. Cambridge University Press, Cambridge.

Goodman S. N. and Berlin J. A. (1994) The use of predicted confidence intervals when planning experiments and the misuse of power when interpreting results. *Annals of Internal Medicine* 121: 200-206.

Hoenig J. M. and Heisey D. M. (2001) The abuse of power: the pervasive fallacy of power calculations for data analysis. *The American Statistician* 55: 19-24.

Hubbard R. and Ryan P. A. (2000) The historical growth of statistical significance testing in psychology - and its future prospects. *Educational and Psychological Measurement* 60: 661-681.

Johnson D. H. (1999) The insignificance of statistical significance testing. *Journal of Wildlife Management* 63: 763-772.

Johnson D. H. (2002) The role of hypothesis testing in wildlife science. *Journal of Wildlife Management* 66: 272-276.

Keough M. J. and Mapstone B. D. (1997) Designing environmental monitoring for pulp mills in Australia. *Water Science and Technology* 35: 397-404.

Lenth R. V. (2000) Java applets for power and sample size. http://www.stat.uiowa.edu/~rlenth/Power/index.html (but regression module currently found at *http://www.stat.uiowa.edu/~rlenth/Power/OldPiFace.html*).

Mapstone B. D. (1995) Scalable decision rules for environmental-impact studies - effect size, Type-I, and Type-II errors. *Ecological Applications* 5: 401-410.

Meehl P. E. (1978) Theoretical risks and tabular asterisks: Sir Karl, Sir Ronald, and the slow progress of soft psychology. *Journal of Consulting and Clinical Psychology* 46: 806-834.

Nelder J. A. (1999) From statistics to statistical science. *Journal of the Royal Statistical Society Series D - The Statistician* 48: 257-267.

Newcombe R. G. and Altman D. G. (2000) Proportions and their differences. In: *Statistics with Confidence* (eds. D. G. Altman, D. Machin, T. Bryant, N. and M. J. Gardner) pp. 45-56. BMJ Books, Bristol.

Neyman J. (1950) *First Course in Probability and Statistics*. Holt, New York.

Neyman J. (1955) The problem of inductive inference. *Communications on Pure and Applied Mathematics* 8: 13-46.

Neyman J. (1957) Inductive behaviour as a basic concept of philosophy of science. *International Statistical Review* 25: 7-22.

Neyman J. and Pearson E. S. (1928) On the use and interpretation of certain test criteria for purposes of statistical inference: Part I. *Biometrika* 20A: 175-240.

Oakes M. W. (1986) *Statistical inference: A commentary for the social and behavioural sciences*. John Wiley and Sons, Chichester.

Osenberg C. W., St Mary C. M., Schmitt R. J., Holbrook S. J., Chesson P. and Byrne B. (2002) Rethinking ecological inference: density dependence in reef fishes. *Ecology Letters* 5: 715-721.

Palmer A. R. (2000) Quasireplication and the contract of error: Lessons from sex ratios, heritabilities and fluctuating asymmetry. *Annual Review of Ecology and Systematics* 31: 441-480.

Pearson E. S. (1962) Some thoughts on statistical inference. *Annals of Mathematical Statistics* 33: 394-403.

Peterman R. M. (1990) Statistical power analysis can improve fisheries research and management. *Canadian Journal of Fisheries and Aquatic Science* 47: 2-15.

Quinn G. P. and Keough M. J. (2002) *Experimental Design and Data Analysis for Biologists*. Cambridge University Press, Cambridge.

Rosnow R. L. and Rosenthal R. (1989) Statistical procedures and the justification of knowledge in psychological science. *American Psychologist* 44: 1276-1284.

Rothman K. J. (2002) *Epidemiology: An Introduction*. Oxford University Press, New York.

Saville D. J. (2003) Basic statistics and the inconsistency of multiple comparison procedures. *Canadian Journal of Experimental Psychology* 57: 167-175.

Schenker N. and Gentleman J. F. (2001) On judging the significance of differences by examining the overlap between confidence intervals. *The American Statistician* 55: 182-186.

Schmidt F. L. (1996) Statistical significance testing and cumulative knowledge in psychology: Implications for training of researchers. *Psychological Methods* 1: 115-129.

Schmidt F. L. and Hunter J. E. (1997) Eight common but false objections to the discontinuation of significance testing in the analysis of research data. In: *What if there were no significance tests?* (ed. J. H. Steiger) pp. 37-64. Lawrence Erlbaum, Mahwah.

Sellke T., Bayarri M. J. and Berger J. O. (2001) Calibration of P-values for testing precise null hypotheses. *American Statistician* 55: 62-71.

Senn S. (1997) *Statistical issues in drug development*. John Wiley and Sons, Chichester.

Smithson M. (2003) *Confidence Intervals*. Sage, Thousand Oaks.

Sokal R. R. and Rohlf F. J. (1995) *Biometry*. W.H. Freeman and Company, New York.

Steidl R. J., Hayes J. P. and Schauber E. (1997) Statistical power analysis in wildlife research. *Journal of Wildlife Management* 61: 270-279.

Steidl R. J. and Thomas L. (2001) Power analysis and experimental design. In: *Design and Analysis of Ecological Experiments* (eds. S. M. Scheiner and J. Gurevitch) pp. 14-36. Oxford University Press, New York.

Stewart-Oaten A. (1995) Rules and judgments in statistics: Three examples. *Ecology* 76: 2001-2009.

Taylor B. L. and Gerrodette T. (1993) The uses of statistical power in conservation biology: The Vaquita and Northern Spotted Owl. *Conservation Biology* 7: 489-500.

Thomas L. and Juanes F. (1996) The importance of statistical power analysis: an example from *Animal Behaviour*. *Animal Behaviour* 52: 856-859.

Tukey J. W. (1977) *Exploratory data analysis*. Addison-Wesley, Reading.

Tversky A. and Kahneman D. (1971) Belief in the law of small numbers. *Psychological Bulletin* 76: 105-110.

Underwood A. J. (1990) Experiments in ecology and management: Their logics, functions and interpretations. *Australian Journal of Ecology* 15: 365-389.

Underwood A. J. (1997) *Experiments in Ecology. Their Logical Design and Interpretation Using Analysis of Variance.* Cambridge University Press, Cambridge.

Wade P. R. (2000) Bayesian methods in conservation biology. *Conservation Biology* 14: 1308-1316.

Wilkinson L. (1999) Statistical methods in psychology journals: Guidelines and explanations. *American Psychologist* 54: 594-604.

Wolfe R. and Hanley J. (2002) If we're so different why do we keep overlapping? When 1 plus 1 doesn't make 2. *Canadian Medical Association Journal* 166: 65-66.

Yoccoz N. G. (1991) Use, overuse, and misuse of significance tests in evolutionary biology and ecology. *Bulletin of the Ecological Society of America* 72: 106-111.

Zar J. H. (1999) *Biostatistical Analysis*. Prentice Hall, Upper Saddle River.

In: New Trends in Ecology Research
Editor: A. R. Burk, pp. 103-117
ISBN: 1-59454-379-8

Chapter 4

SOME ASPECTS OF COMPUTER TECHNOLOGIES APPLICATION TO ECOLOGICAL MODELLING: CASE STUDIES OF MICROSOFT EXCEL, AND 'RELIEF-PROCESSOR' - A MODELLING SYSTEM FOR GEOECOLOGICAL RESEARCH

V. Krivtsov[1], S. Kostrikov[2], H.J Staines[3], B. Vorobiov[2] and A. Brendler[3]

[1]School of Ocean Sciences, University of Wales Bangor,
Marine Science Laboratories, Menai Bridge, Anglesey LL59 5AB, UK
[2]Geography Department, Kharkov State University,
Maidan Svobodu 4, Kharkov 310077, Ukraine
[3]SIMBIOS, Schools of Science and Engineering and Computing,
University of Abertay Dundee, Bell Street, Dundee DD1 1HG, Scotland, UK

ABSTRACT

The correct understanding of ecological and environmental processes is indispensable for the sustainable development of Humankind. For practical purposes, the account of ecosystem complexity is greatly aided by the application of computer aided technologies. This paper presents a methodological approach linking computer modelling of natural processes with geoinformation technology, realised in the PC modelling System "Relief-Processor" (R_P). The paper also delivers a word of warning as regards possible pitfalls with application of certain computer technologies for modelling of ecological processes, giving an example of errors which may occur while using Microsoft Excel for fitting trendline equations.

The methods of landscape assessment implemented in the R_P are based on the complex numerical algorithms of structural analysis and allow not only generation of the information missing in the initial data sets, but also calculation of various geomorphological-hydrological indexes, representing current conditions of the

[1] Email for correspondence: e96kri69@netscape.net or oss00d@bangor.ac.uk

environment. This research technique is shown to facilitate solution of various environmental problems, including control of sediment-pollutant delivery from both point and non-point sources, allocation of water-supply reservoirs in arid landscapes, modelling of channel network, investigation of erosion processes /sediment yield-delivery, studies of landscape geochemistry and pollutant transgression, etc. Coupling of the core modules with specific detailed simulation ecological models further widens the scope of the methodology presented here.

Implementation of the advanced modelling techniques, facilitating both the comprehension of complexity of "natural landscape - human activity interactions" and the evaluation of economic consequences of various unfavourable environmental events, accuracy, precision and high quality of cartographic output, are the main conditions necessary for the evolution of any geoinformation system into a real Decision Support System. The software presented here meets all these requirements, and should, therefore, prove useful in Landscape and Ecological Engineering, Land Management issues, Environmental Assesment and Auditing.

Key words: Ecological modelling, GIS, relief modelling, watershed morphology, hydrology, landscape architecture, structural analysis, pollutant pathways, quadratic model, linear equation, polynomial regression, category charts, scatter plots

INTRODUCTION

The correct understanding of the ecological and environmental processes is indispensable for the sustainable development of Humankind. For practical purposes, the account of ecosystem complexity is greatly aided by the application of computer aided technologies, such as GIS, simulation models, expert systems, etc. Here we consider the following examples of some aspects of advantages and disadvantages associated with the application of computer technologies. Firstly, we outline a methodological approach linking computer modelling of natural processes with geoinformation technology, realised in the PC modelling System "Relief-Processor" (R_P). Secondly, the paper also delivers a word of warning as regards possible pitfalls with application of certain computer technologies for modelling of ecological processes, giving an example of errors which may occur while using Microsoft Excel for fitting trendlines for fungal biomass dynamics.

CASE STUDY 1: GIS MODELLING REALM FOR GEOECOLOGICAL RESEARCH PURPOSES - IMPLEMENTATION IN THE PC MODELLING SYSTEM "RELIEF-PROCESSOR"

Although the need for coupling advantages provided by geoinformation systems with various ecological submodels is well recognised, there are only a few hitherto developed environmental simulation modelling systems that have been elaborated either as inner GIS modules or as external tools. All these systems, to a greater or lesser degree, use either deterministic or stochastic approaches. Geographical stochastic environmental models, beginning with Shreve's model (Shreve, 1966), are mainly based on the calculation of natural events and process probabilities. The systems incorporating environmental deterministic simulation models are normally characterised by a thorough choice of functional interdependencies/parameters describing interrelations among physical components of the landscape and a solution procedure for differential equations (since the equations describing environmental systems are characterised by a very high order and strong non-linearity). It is evident that both of these somewhat unilateral approaches may distort some important characteristics of the research area. Here, we present an approach capable of minimising such distortions by modelling the appropriate information missing in an initial data set, using a so-called "transitional research subject" with consideration of both geomorphological and hydrological aspects of the environmental domain.

Transitional Research Subject

It is obvious that geomorphological processes and hydrological events on river watersheds are the most crucial among all significant environmental processes responsible for the spatial organisation of a landscape. The current landform morphology acts as a dominant factor in the determination of water flow/sediment/pollutant delivery directions, and may therefore be used as a key to the current geomorphological-hydrological regionalisation. Hence the cause-result sequence within water watershed boundaries could be defined as "geomorphological processes - hydrological responses - landscape evolution".

Since the two main components of fluvial topography (i.e. channel network and catchment morphology) are related to the fundamentals of a runoff process, it seems logical to assume that both of them may be determined by the overland flow. For a modelling tool such assumptions incur a great number of parameters and variables that have to be simulated, e.g. topological-geometrical properties of the channel network, interactions between streams of a different order, etc. It is worth pointing out that the dynamic nature of the relationships among the environmental factors determining conditions of the landscape lead to continuos changes in the latter in space and time. Consideration of this dynamics is an important feature of the proposed approach, reflected in the definition of the modelling research subject as "transitional".

The assessment of catchment morphology by an appropriate modelling tool, accounting for the transience of the research subject, can lead to some considerations and conclusions not only about fixed parameters of a landscape's spatial organisation, but also about dynamic processes, important for the functioning of the considered landscape. For instance, the relationship between average discharges and threshold discharges required to disrupt initial channels (first-order streams – see Strahler, 1952), may be determined as a modelling output.

The watershed of an initial channel could be regarded as an important environmental indicator, because the transformation of sheet runoff into channel flow mirrors the regional physioclimatic picture and strictly depends upon the conditions of a particular landscape. Thus, an interactive definition (within the modelling simulation shell) of an initial channel watershed area is an indispensable step in the proposed approach.

For a successful modelling outcome, however, some other landscape features should also be taken into account. For instance, sediment delivery rates are directly dependent on both soil loss rate and transport efficiency of surface runoff channel flow. Soil loss can be estimated from the average characteristics in terms of watershed slopes, representative vegetation and the erodibility factors. Thus, the scope of modelling research becomes broader than geomorphological-hydrological only - it emanates as being truly environmental. This predestination of a GIS-modelling realm corresponds to the existing definition of GIS as applications which represent spatial-temporal aspects of natural objects in "a flexible and continuous way" (Asproth, 1993).

MODELLING METHODOLOGY

After selecting a number of key processes among all the processes evolving in a natural ecosystem, one has to designate the resulting consequences of these processes as the main subject of the research. Assuming that the present landform morphology in fluvial humid regions performs in strict relation with a number of other landscape factors and components (Richards, 1986), it has to be considered as a key factor not only in hydrological, but also in environmental (i.e. physico - geographical) regionalisation.

Spatial landform conjunctions and other morphological peculiarities determine main runoff directions, which, in turn, determine routes of riverine loading of various pollutants (nutrients, heavy metals, anthropogenic organic compounds, etc.). Thus, for consideration of pronounced effects of riverine loading on most ecosystems, examination of geomorphological-hydrological aspects of a natural landscape becomes indispensable. The latter subjects constitute the scope of landscape ecology or, in other terms, - geoecology. A geoecological approach combines two kinds of spatial procedures - both geomorphological-hydrological and environmental analysis. Elementary landscapes are studied at the geoecosystem level characterised by biogeocoenosis with steady trophic webs and biogeochemical/energy flow patterns. It is reasonable to conduct such research within the boundaries of a river watershed (Kostrikov, 1989). In this case a spatial modelling subject is adapted to its natural object, because watershed boundaries represent the direct result of definite conformities to natural laws manifested in spatial location of fluvial landforms (Abrahams, 1975). Hence, the watershed research approach may be regarded as the first principle of our main conceptual model for the GIS modelling realm, while the structural method is the second one.

Well known dependencies between the earth surface (fluvial) landforms and the functional structure of an elementary ecosystem give the opportunity to trace any changes in a natural landscape affected by anthropogenic activities. To complete this task, however, the spatial variation of the natural landscape should be accounted for. Thus the structure of the earth surface topography (which is determined as the particular regional arrangement of topographic parts, elements, and constituents of the landscape (Kostrikov, 1987)) should be investigated applying the principles mentioned above. It is worth pointing out that the structural method is limited by the omission of geological control and genetic factors. Despite that, it allows determination of route network of sediment delivery and pollutant distribution according to two kinds of invariant lines within the relief (i.e. channel network lines and watershed net lines).

Regarding the construction of the previously mentioned conceptual model, investigation of earth surface topography must examine variation in channel (river-gully) network as the first step (Kostrikov, 1989). It is done for the purpose of extraction of process-related information from channel network patterns and/or from their related geomorphological indexes. Since earth surface topography is one of the main geographical landscape determiners (it redistributes runoff and channel flow, input of solar energy and has a direct influence on erosion processes) (Chervaniov, 1982), each channel network pattern (as a set of fluvial elements) has governed the ecological balance of the given territory in its own (network pattern's) way. The relief structure, therefore, should be defined as the structure of a geographical landscape and the spatial hierarchical organisation of the relief should be considered among important factors determining the landscape development.

Conventionally, fluvial geomorphology has used topographical decomposition to divide the landscape into its components - landforms (Horton, 1945, Shtraler, 1952). Hence, topography (relief) may be represented as a hierarchical system, comprising a number of landforms. Having combined relief structural analysis with the catchment (watershed) approach to modelling research, one gets an opportunity to describe the fluvial relief in terms of the complex system of subcatchments (i.e. smaller parts of the watershed of a higher order). Subcatchments in such a system should fit together within the boundaries of a higher order watershed in respect of their sizes, shape and migration rates of valley bends, as well as the length and steepness of the main channels. The channel network could be regarded as a multidimensional continuum with overlapping and transitional characteristics, reflecting the diverse environmental controls of climate, hydrology, sediment supply, proximal-distal location of topographic elements, watershed tectonics and pollutant yield/delivery. After the channel network is defined, estimation of the contamination at a given point within the watershed boundaries is relatively simple. For instance, an area of the affiliated surface-channel flow can be estimated for any point in the catchment as S1. Then the area of intersection between a contaminating source area (S2) and an area of flow at the requested point could be calculated as S1/S2. The latter expression will determine the level of pollution concentration at the considered point.

Having modelled geochemical migration processes, we introduce the notion of "structural transfer of pollutants" as a measure of influence of the channel network on the upland erosion and geochemical transit within the landscape. Within such an approach it is possible to estimate the length and force of pollution migration using various characteristics of the catchment and substances involved. This helps to overcome a number of serious problems normally encountered in the estimation of pollution transgression (for details see Kostrikov and Vorobiov, 1998).

Brief Description of the PC Modelling System "Relief-Processor"

Clearly, any universal environmental GIS should contain and properly handle both the topographic and channel network information, and be able, therefore, to account for the diverse environmental controls of climate, hydrology, sediment supply, proximal/distal channel location, as well as lithological and geological-tectonic factors. The simulation-modelling approach of such a system should satisfy the strict cartographic requirements. Data planned for constant informative storage should reflect the structural peculiarities of topography and possible directions of its changes due to the influence of water related processes, weathering, soil creeping and mass movement. The mathematical topographical model should be based on the determined geomorphological structural model, accounting for information related to channel network and catchment morphology. This forms the necessary prerequisites for formalisation and the strict description of the relief structure in a mathematical language (Vorobiov et al, 1995).

The automated modelling system "Relief-Processor" (R_P) applies the approach presented in this paper to representation, processing, and storage of earth surface topography information and to some other natural environmental data (Vorobiov and Kostrikov, 1992) and does therefore satisfy all the requirements mentioned above. Currently the basic components of the R_P system are SPATIAL DATA ANALYSIS, STRUCTURAL ANALYSIS, MORPHOMETRY and CARTOGRAPHY. In addition to these subsystems the R_P contains such useful auxiliary utilities as graphic editors built in the Structural Analysis and Cartography subsystems and a network calculator included in each subsystem. The conceptual framework for R_P advanced geographical applications foresees the creation of new main information domains in addition to the existing subsystems mentioned above. It is worth pointing out that the approach used is open to adjustment and the software structure could readily incorporate sub-models describing detailed dynamics of any geoecosystem components. For example, linking of the R_P with a dynamic lake ecosystem model Rostherne (Krivtsov et al., 1998, 1999, 2000) is currently considered as a potential extension of our research.

All control of the R_P is performed through menus, dialogue boxes and hot-keys. All implemented functions and operations can be activated either with keys or with a mouse. The interface has the following types of controls: sensor bars, pop-up menus, selection lists, input fields and windows, lists with input fields and dialogue boxes.

The subsystem SPATIAL DATA ANALYSIS is designed for data input and their preprocessing for the purposes of further research. It offers a wide range of approximation and interpolation algorithms as well as cluster-analysis. This subsystem normally operates with a .DAT extension ASCII data files of an ordinary text data format. Input files are initially prepared in a standard xyz format and may contain either regular or irregular grids. Generally speaking, any scalar field presented on a plane may be regarded as a research subject for this computer package.

Input files must be processed at the first stage of data manipulation. It is equally possible to treat an input data file as a whole, or deal with some part of the input grid, but in any case data correctness is completely up to a user. An initial ASCII-file is converted in this part of the R_P into a binary computer code file with a .NET extension. At this stage, providing the grid is regular and the input information has been entered correctly, the regional boundaries are traced and stored in a .NET file after conversion.

The possibility of processing the data obtained in an irregular grid is the crucial advantage of the R_P. Regularisation is provided by Approximation, Triangulation, and Criging algorithms (e.g. Geographical Information Systems and Cartographic Modeling, 1990). When a .NET output file is successfully created, the user can "sew-up" several adjacent grids in a single unit. The latter possibility is especially useful for processing of irregular data, since even a modest increase in the amount of data handled by the regularisation procedure results in a dramatic increase in the amount of potentially useful information wasted (Bailey, 1992). Linking the grids together is done by the Net Linking operation, which normally completes the work of the subsystem. This operation also includes search for boundaries, filtration and creation of the . NET output file for further processing.

The subsystem STRUCTURAL ANALYSIS is the most authentic part of the R_P based on the original structural approach discussed above. This subsystem is the core of the whole package implementing our original research approach. It allows on the basis of only topographic data and some initial hydrological parameters delineation of both the complete erosion structure in boundaries of any fluvial watershed (or their set) and a complete network of talwegs and watershed boundaries of any order.

At the first step within this subsystem the definition of relief structural lines (networks of channels and drainage divisions) is performed. This allows modelling topological and geometric properties of fluvial topography, helping to reveal structural laws of relief self-organisation and regulation.

At the next (optional) step the subsystem allows simulation of different types of human impact on natural landscape within catchment boundaries (e.g. flooding of a water reservoir after building a dam, modification of natural relief by mine construction, transfer of pollutants from both point and non-point sources, increase in the sediment yield-delivery as a result of alterations in natural relief erosion structure, etc.). Naturally, such simulations require a number of environmental variables and parameters as input information, such as area of initial watershed, runoff ratio, default level of contamination in the region, etc. Model simulations can be easily compared with empirical observations. A number of report (TXT) files, describing different fluvial relief morphometric characteristics as well as all the related environmental information, can be optionally generated at this stage as an output. To facilitate further data handling the output of the subsystem is saved in a special format (files with .STR and .RED extensions intended for further processing are generated by the subsystem automatically).

Morphometry

The subsystem Morphometry gives an opportunity to create on the basis of a .NET file containing the initial topographic information a set of .NET files with secondary parameters. For instance, if the primary file contains the local relief elevation values, then the user may obtain relief gradients, azimuths of topographic elements and many other morphometric characteristics. Moreover, if the structural research has already been done, it is also possible to calculate (for any point of the grid) such indexes as velocities and powers of surface runoff, as well as to construct cut-in surfaces and other structural characteristics of the relief.

A set of obtained morphometric grids (slopes, azimuths, bends of gradients, etc.) allows the tracing of complex earth surface processes. After a successful calculation, output and storage of the obtained information, it is usually necessary to use a provided filter option. The latter allows creation of a more attractive map of the calculated network by smoothing the distortions inevitably introduced by the numerical algorithms used for data processing.

The main objectives of the subsystem CARTOGRAPHY are visualisation, generalisation and the generation of a user-friendly overview of all the research done within other subsystems. Here results obtained are presented as isolined and shaded maps (two or three dimensional) or as some other highly attractive images. Using current advanced photo plotters as a hard-copy device, it is possible to obtain colourful high quality pictures and overview three-dimensional graphics that could be variously applied in landscape architecture and other subject fields.

APPLICATIONS IN PARTICULAR GEOECOLOGICAL DOMAINS

The R_P modelling system has already been applied to quite a few major problems in several geographic regions, including structuring of a channel network and delivery of the sediment yield originated due to the erosion processes; landscape geochemistry and pollutant delivery; allocation of water-supply reservoirs in arid landscapes (the Crimean region). For the purposes of this paper only a few brief examples will be given.

Example I

Studying by means of the Relief-Processor package the development of a gully network in a non-steady state on the basis of extensive data sets describing Ukrainian river watersheds of different orders, we have estimated the drainage system dominated by ephemeral flow and the erosion/sediment delivery process controlled by catchment morphology and sediment availability, and timing and characteristics of hydrological events. The software clearly indicated that the further the gully network is from the equilibrium, the further from this state remains the whole natural landscape, no matter whether the gully growth was initiated by human impact or by hydrological events. This example not only describes a highly successful application to geomorphological research, but also demonstrates usage of the software as a useful tool for environmental appraisal (Kostrikov, 1989).

Example II

The comparison of modelling results with the sets of empirical data from the Noth-Eastern Ukraine concerning fluvial landforms and channel networks, suggests the following conclusions related to the different kinds of an anthropogenic impact:

1. Channel network and catchment morphology depend upon land use only indirectly. However, agriculture, surface mining and building site construction dramatically change the relief, enhance erosion, expose and disrupt the ground surface. In turn, as it has also been shown by Abrahams (1975), the latter processes keep generating new channels until the network finally attains a steady state.
2. Within areas of similar physiography, topography and climate, sediment-pollutant supply is controlled primarily by land use, and secondarily by landscape conditions.
3. Sediment delivery is controlled principally by local physiography and secondarily by land use.

Example III

The problem of water-supply has been rather acute in the Crimean peninsula for the last two decades because of local arid climate and growth in water consumption. The research previously conducted in this region helped to alleviate the problem only marginally. Using R_P, however, even the ephemeral channel network can be modelled fairly well with the computation of the principal components of the water balance for the designed reservoirs.

Also the package allowed evaluation of quite a few hydrometeorological components of the local landscape, leading, therefore, to the solution of the interaction problem among the components of the water balance. In summary, the research conducted in Eastern Crimea and on the Chiornaya river-watershed achieved the following goals:

1. A proposed definition of water resources.
2. Quantification of the water resources of designed reservoirs.
3. Identification of types of probable erosion processes (as results of water flow regulation by reservoirs).

The proposed methodology, therefore, proved applicable to cases requiring evaluation of the effect of reservoirs on local water resources and landscape processes.

The approach introduced and the software presented here have also been used sucessfully as decision support methodology and an automated decision support tool respectively in the case study of wastes originated from an iron ore plant.

CONCLUSION

Generation of environmental information unobtainable during the initial information arrangement (field research, database elaboration, etc.) is an important procedure within the GIS modelling realm. The consequent steps in a modelling process based on the structural approach are as follows: surface runoff model = > vectorisation of catchment areas and generalisation of erosion structure = > hierarchical model of hydrographical network in relation with catchment bounds = > coupling with various ecological submodels and with submodels of pollutant transformation.

All the points mentioned above represent a synopsis of our modelling methodology intended for precise elaboration of the information concerning the nature and relative significance of environmental processes and their characteristics within the natural landscape. The proposed research technique implemented in the R_P modelling package advocates moving from a conventional GIS towards the Environmental Decision Support System (EDSS), satisfying the methodological requirements for "integrating the GIS technology with computer based Decision Support Systems" (Grothe and Scholten, 1992).

The results of the research reported above demonstrate that the modelling approach presented here could provide meaningful and consistent estimation not only for geomorphological, but also for a much broader set of environmental parameters, satisfying, therefore, the requirements for the GIS modelling realm. In our view this approach is within the trend of those information systems that have led to the design of advanced GIS (Scholten and others, 1994). Combining the described modelling system with an appropriate GIS shell and various ecological submodels promotes the elaboration of a geoinformation system capable of describing the complex and cyclic response of a natural landscape to the superfluous anthropogenic activity.

Case Study 2: Have You Fitted the Right Trend Line? - A Warning to Fellow Environmental Scientists who Over-Rely on Microsoft Excel

Microsoft Excel is a commonly available software, and it is preferred by many scientists due to its user-friendly interface, clear instructions, versatile help system, good graphics, compatibility with other commonly used computer products, and availability of database tools, statistical functions, and a powerful built-in computer language – Visual Basic for Applications (VBA – for details of the excellent potential and pitfalls of the latter see Kofler, 2003, and references therein). In particular, the software is popular among environmental and natural scientists, and the authors have variously used it, along with other software, for data handling and analysis in studies related to environmental technology (Krivtsov et al. 2004) protozoology (Krivtsov et al. 1999c), poultry science (Hazary et al. 2000, 2001; Staines et al. 1998; Wishart and Staines 1995; Wishart et al. 2001), biogeochemistry (Krivtsov et al. 2000b; Krivtsov et al. 2001b), algology (Krivtsov et al. 2000c; Krivtsov et al. 2002b), environmental chemistry (Krivtsov et al. 2001c; Krivtsov et al. 2002a), forest science (Srinivasan et al. 1992), animal ecology (Albon et al. 1992), and ecological and environmental modelling (Krivtsov et al. 1998; Krivtsov et al. 1999a; Krivtsov et al. 1999b; Krivtsov et al. 2000a; Krivtsov et al. 2000d; Krivtsov 2001; Krivtsov et al. 2001a; Krivtsov et al. 2004; Krivtsov et al. 2004))

Modelling using linear or higher order polynomial regression is widely used, e.g. in natural sciences for making calibration curves and describing relationships between physical, chemical, physiological and environmental variables (Draper and Smith 1998). Consequently, various commercial packages, including Minitab, SAS, SPSS, and Microsoft Excel can be used to find the least squares estimates of parameters (coefficients) in the model. In Microsoft Excel this procedure is implemented by displaying the equation of a trendline fitted to a dataset graphed on a chart. This may be a very useful procedure, and we routinely use it in data exploration and analysis (e.g. Krivtsov et al., 2002a; 2001b). However, here we show that the results of this procedure, in certain cases, may be incorrect, and should, therefore, be treated with caution.

Results and Discussion

To illustrate a problem which may occur while fitting a trend line in Microsoft Excel, consider modelling the concentration of commercial Bovine Serum Albumin standards, assayed using the Bradford method - routinely used for protein determinations (Bradford 1976) - by the 595 nm absorbance values (the authors used it to estimate the soil level of glomalin, a protein which could be used as a biomarker for arbuscular mycorrhizal fungi). The observed data are shown in the first 2 columns of Table 1 whilst Figure 1a shows a scatter plot with the fitted values from the linear and quadratic models. The parameter estimates are identical with those obtained by other (e.g. SPSS, Minitab) commercial software.

Fig. 1a

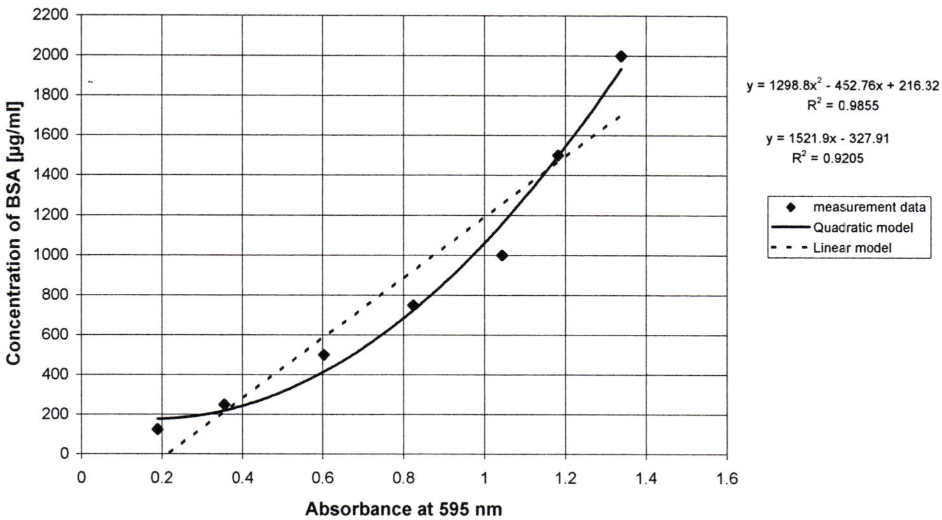

Fig. 1b

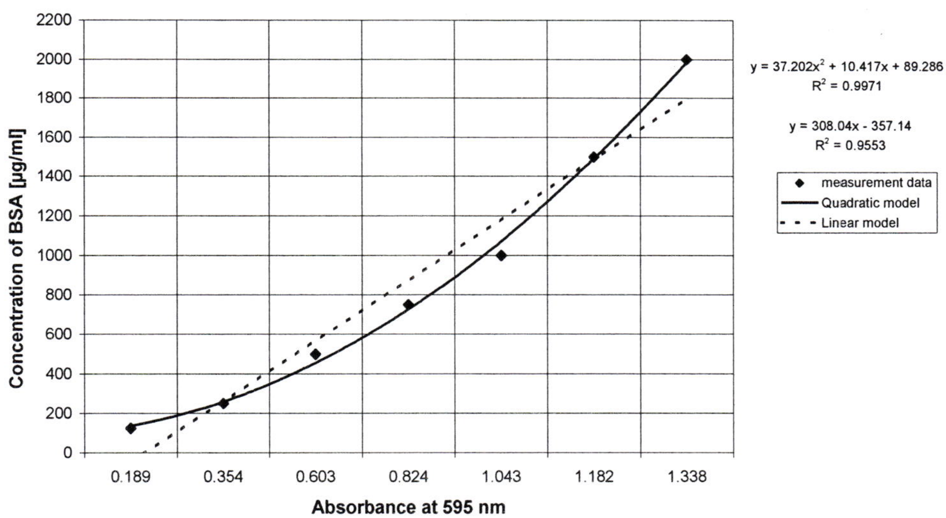

Figure 1. Approximation of the relationship between BSA concentration and 595 nm absorbance values using a linear and a quadratic function is displayed as a scatter (upper panel - Fig. 1a) and a line (lower panel - Fig. 1b) plot. Note that coefficients of the equations fitted using the scatter plot (upper panel) are correct, while the coefficients of the equations fitted using the line plot (lower panel) appear to be incorrect (see text for further commentary).

If, however, the type of the Excel chart is changed to a line plot, the fitted values from both models are identical to Figure 1a but the coefficients of the models change dramatically (Figure 1b). The problem arises since in the latter case the horizontal axis is treated as a category axis, and the abscissa values used to fit the equations merely represent the ordinal numbers of the subsequent categories (see the last column of Table 1). Hence, the specific values of the absorbance (column 2, Table 1) are simply displayed as labels, and not used in the calculations. The resulting models are, therefore, fitted to the data set comprising Columns 1 (dependent variable) and 3 (independent variable) of Table 1.

Table 1. The dataset used to create charts displayed in Figure 1. Values of absorbance at 595 nm are given for the corresponding concentrations of Bovine Serum Albumin (BSA) standards made up from a commercially purchased stock (Pierce Chemical Company) and assayed by the Bradford method

Concentration of BSA [μg/ ml]	Absorbance at 595 nm	Category Number
125	0.189	1
250	0.354	2
500	0.603	3
750	0.824	4
1000	1.043	5
1500	1.182	6
2000	1.338	7

Predicting BSA concentration values for an absorbance of 1 yields values of 1194 and 1062 for the correct linear and quadratic models respectively. However, predictions of –49 and 137 are obtained from the linear and quadratic models respectively using the incorrect parameter estimates.

Therefore, although Excel Scatter plots seem to be an appropriate choice for data approximation using trendline equations, application of category charts for this purpose appears to be inappropriate, and may lead to gross prediction errors. This should be borne in mind by any scientist using Microsoft Excel for their work. No software is fully versatile, and even the common commercially available packages are, in many respects, not always foolproof.

REFERENCES

Abrahams, A.D. (1975) Topological random channel network in the presence of environmental controls. *In: Geol. Soc. Am. Bull.*, Vol.86, P. 1459-1462. Publisher: NY.

Albon S. D., Staines H. J., Guinness F. E. and Cluttonbrock T. H. (1992) Density-Dependent Changes in the Spacing Behavior of Female Kin in Red Deer. *Journal of Animal Ecology* 61: 131-137.

Asproth, V. (1993) GIS, Geo Informatic Systems Applications. In: MicroCAD'93, *International Computer Science Meeting*, Hungary. Budapest/Miskolc.

Bailey, T.C. (1992) Statistical Spatial Analysis and Geographical Information Systems. A review of the potential and progress in the srtate of the art. *Proceedings of Third*

Bradford M. M. (1976) A rapid and sensitive method for the quantitation of microgram quantities of protein utilizing the principle of protein-dye binding. *Anal. Biochem.* 72: 248 - 254.

Chervaniov, I.G. (1983) Relief structure and its influence on landscape structure. *(In Russian). Physical Geography and Geomorphology,* 30, 104-108.

Draper N. and Smith H. (1998) Applied Regression Analysis. John Wiley, New York.

European conference of GIS, edited by H. Ottens, 1, 186-203.

Geographical Information Systems and Cartographic Modeling. (1990) *Englewood Cliffs, NJ*: Prentice Hall.

Grothe M., Scholten H. (1992) Modelling catchment areas, towards the development of spatial decision support systems for facility location problems. In: *Third European Conference of GIS*, II, 978-987.

Hazary R. C., Staines H. J. and Wishart G. J. (2000) Assessing the efficiency of mating in broiler breeder flocks by enumerating the spermatozoa which penetrate the inner perivitelline layer over the germinal disc. *British Poultry Science* 41: 395-400.

Hazary R. C., Staines H. J. and Wishart G. J. (2001) Assessing the effect of mating ratio on broiler breeder performance by quantifying sperm : egg interaction. *Journal of Applied Poultry Research* 10: 1-4.

Horton R.E. (1945) Erosional Development of Streams and Their Drainage Watersheds: Hydrological Approach to Quantitative Morphology. *Bulletin of American Geological Society,* 56, 75-370.

Kofler, M. (2003). Definitive guide to Excel VBA (2 edition). Springer Verlag, Heidelberg.

Kostrikov S.V. (1987) Experience of analysis of relief horizontal structure (in Russian). *Physical Geography and Geomorphology,* 34, 123-129.

Kostrikov S.V. (1989), Analysis of structural network of Oskol-river watershed for studying of erosion processes (in Russian). *Physical Geography and Geomorphology*, 36, 87-93.

Kostrikov, S.V. and B.N. Vorobiov. 1998. Landform-dependent factor of landscape geochemical migration and possibility of its formalised definition (*in Ukrainian*). *Ukrainskiy geographicheskiy zgurnal.* 3, 23-29.

Krivtsov V., Bellinger E., Sigee D. and Corliss J. (1998) Application of SEM XRMA data to lake ecosystem modelling. *Ecological Modelling* 113: 95-123.

Krivtsov V., Bellinger E. G. and Sigee D. C. (1999a) Modelling of elemental associations in Anabaena. *Hydrobiologia* 414: 77-83.

Krivtsov V., Sigee D., Corliss J. and Bellinger E. (1999b) Examination of the phytoplankton of Rostherne Mere using a simulation mathematical model. *Hydrobiologia* 414: 71-76.

Krivtsov V., Tien C., Sigee D. and Bellinger E. (1999c) X-ray microanalytical study of the protozoan Ceratium hirundinella from Rostherne Mere (Cheshire, UK): Dynamics of intracellular elemental concentrations, correlations and implications for overall ecosystem functioning. *Netherlands Journal of Zoology* 49: 263-274.

Krivtsov V., Bellinger E. and Sigee D. (2000a) Incorporation of the intracellular elemental correlation pattern into simulation models of phytoplankton uptake and population dynamics. *Journal of Applied Phycology* 12: 453-459.

Krivtsov V., Bellinger E., Sigee D. and Corliss J. (2000b) Interrelations between Si and P biogeochemical cycles - a new approach to the solution of the eutrophication problem. *Hydrological Processes* 14: 283-295.

Krivtsov V., Bellinger E. G. and Sigee D. C. (2000c) Changes in the elemental composition of Asterionella formosa during the diatom spring bloom. *Journal of Plankton Research* 22: 169-184.

Krivtsov V., Corliss J., Bellinger E. and Sigee D. (2000d) Indirect regulation rule for consecutive stages of ecological succession. *Ecological Modelling* 133: 73-82.

Krivtsov V. (2001) Study of cause-and-effect relationships in the formation of biocenoses: Their use for the control of eutrophication. *Russian Journal of Ecology* 32: 230-234.

Krivtsov V., Goldspink C., Sigee D. C. and Bellinger E. G. (2001a) Expansion of the model 'Rostherne' for fish and zooplankton: role of top-down effects in modifying the prevailing pattern of ecosystem functioning. *Ecological Modelling* 138: 153-171.

Krivtsov V., Sigee D. and Bellinger E. (2001b) A one-year study of the Rostherne Mere ecosystem: seasonal dynamics of water chemistry, plankton, internal nutrient release, and implications for long-term trophic status and overall functioning of the lake. *Hydrological Processes* 15: 1489-1506.

Krivtsov V., Sigee D. C., Bellinger E. G. and Porteous G. (2001c) Determination of P release from Rostherne Mere sediment cores. *Acta Hydrochimica Et Hydrobiologica* 29: 111-117.

Krivtsov V., Bellinger E. and Sigee D. (2002a) Water and nutrient budgeting of Rostherne Mere, Cheshire, UK. *Nordic Hydrology* 33: 391-414.

Krivtsov V., Sigee D. and Bellinger E. (2002b) Elemental concentrations and correlations in winter micropopulations of Stephanodiscus rotula: an autecological study over a period of cell size reduction and restoration. *European Journal of Phycology* 37: 27-35.

Krivtsov V., Wäger P. A., Dacombe P., Gilgen P. W., Heaven S., Hilty L. M. and Banks C. J. (2004) Analysis of energy footprints associated with recycling of glass and plastic-case studies for industrial ecology. *Ecological Modelling* 174: 175-189.

Krivtsov V., Walker S. J. J., Burt-Smith G., Staines H., Watling R. and Garside A. (2004) Integrative analysis of ecological patterns in an untended temperate woodland utilising standard and customised software. *Environmental Modelling and Software* 19: 325-335.

Richards K. (1986) Fluvial Geomorphology. *Progress in Physical Geography*. 10/3, 401-420.

Scholten H., Bogaerts H., Verspuy M. (1994). Towards a spatial meta information system of Europe. *In: GIS in Business'94*, Amsterdam. Longman Group UK Ltd. Cambridge. 201-207.

Shreve R.L. (1966) Statistical law of stream numbers. *Journal of Geology*, 74, 17-37.

Srinivasan U., Staines H. J. and Bruce A. (1992) Influence of Media Type on Antagonistic Modes of Trichoderma Spp against Wood Decay Basidiomycetes. *Material Und Organismen* 27: 301-321.

Staines H. J., Middleton R. C., Laughlin K. F. and Wishart G. J. (1998) Quantification of a sperm-egg interaction for estimating the mating efficiency of broiler breeder flocks. *British Poultry Science* 39: 273-277.

Strahler, A.N. (1952) Hypsometric (area-altitude) analysis of erosional topography. *Bulletin of the American Geological Society*, 63, 1117-1142.

Vorobiov B.N., Alesinsky A.I., Kostrikov S.V., Petrenko A.L. (1995) The object- oriented approach to the making of GIS as a base for creation of decision support systems. *Proceedings of the First Joint European Conference and Exhibition of Geographical Information*. 428-430.

Vorobiov B.N., Kostrikov S.V. (1992) The Structural Approach to the Making Topographic GIS: Main Ideas and Realisation. *Third European Conference of GIS*, II, 1304-1313.

Wishart G. J. and Staines H. J. (1995) Assessing the Breeding Efficiency of Broiler Breeder Flocks by Measuring Sperm Transfer into Laid Eggs. *British Poultry Science* 36: 317-323.

Wishart G. J., Staines H. J. and Hazary R. C. (2001) Evaluation of fertility: biological basis and practical application. *Worlds Poultry Science Journal* 57: 309-314.

In: New Trends in Ecology Research
Editor: A. R. Burk, pp. 119-134
ISBN: 1-59454-379-8

Chapter 5

CHANGING CONSUMER-SPECIFIC ECOLOGICAL CONSCIOUSNESS: THEORY AND EVIDENCE FROM SURVEY DATA

Nikolaos Georgantzís**, *Aurora García-Gallego* and *Gerardo Sabater-Grande
Department of Economics and *LEE/LINEEX*
Universitat Jaume I Castellón-Spain

ABSTRACT

At a simple glance, an increase in the consumer's willingness to pay (WTP) for an improvement in the ecological performance of a marketed product should be welfare improving. Furthermore, it should create the appropriate incentives for firms to get involved in such improvements by benefiting "green" producers and harming ecologically indifferent ones. We provide some theoretical arguments against these apparently obvious affirmations, showing that the profitability and welfare effects of WTP increases for green products depend on whether they are targeted towards the least ecologically aware of the consumers. Our hypothesis concerning these different kinds of changes in ecological consciousness is supported by Spanish survey data on subjective perceptions of the consequences of ecological consciousness-increasing advertisements. The results show that, according to our respondents, the socially undesirable type of messages represents a 40% of the total, calling for a more cautious attitude towards qualitative aspects of state advertising aimed at creating a more ecologically aware consumer.

Key words: WTP changes, ecological products, environmental consciousness
JEL: A13, L13, D43, D62

* The authors can be contacted at: Departamento de Economía, Universitat Jaume I, Campus Riu Sec, 12071-Castellón-Spain. Tel: 0034-964-728588. Fax: 0034-964-728591. E-mail: georgant@eco.uji.es.

INTRODUCTION

In recent years, consumers have become increasingly aware of the problems associated with the environmental externalities resulting from the production and consumption of marketed products. It is commonly accepted as obvious that such an increasing consciousness of the consumers towards environmental issues is socially beneficial. Furthermore, a natural consequence of this apparently obvious principle is that by increasing the consumer's consciousness and, thus, her willingness to pay for an improvement in a product's environmental performance, we necessarily create the right incentives for firms to invest in such improvements.

In this chapter, we present survey data on people's perception of the effects caused by state advertising campaigns aimed at increasing the consumer's awareness towards environmental issues. In fact, we distinguish between messages affecting more the willingness to pay for a more environmental friendly product of the most environmentally conscious consumers and messages affecting more the least conscious ones. Results indicate that both types of messages seem to exist and that the former could be more frequent than the latter.

We propose a theoretical framework in which both market and environmental considerations are taken into account. In very general terms, our findings indicate that increasing the ecological consciousness of the *least* ecologically-oriented consumers is socially superior as compared to messages targeting the *most* ecologically oriented ones. In most cases, private profitability is incompatible with social welfare maximization. That is, the most beneficial type of message is the least likely to be supported by privately funded advertising. State campaigns should fill the gap, by undertaking the task of funding messages targeting the least ecologically conscious consumers. Unfortunately, our survey data on subjective perceptions of similar campaigns in Spain indicate that the contrary has been achieved in 40% of the cases.

This chapter is organised in the following way: Sections 2 and 3 present theoretical analysis and predictions. In Section 4, we present and discuss data from a survey. Section 5 concludes.

THEORETICAL CONSIDERATIONS

Consumers are most often heterogeneous regarding their valuations of certain product characteristics[1], as has been already confirmed for most other products of a *public good* nature.[2] Furthermore, the results by Camacho et al. (2003, 2004) extend this finding to private

[1] Due, for example, to different levels of homo economicus and homo politicus in their utility functions. See Nyborg (2000).

[2] Generally speaking, most of the studies using the contingent valuation method to estimate a population's WTP for a private or a public good, implicitly assume some heterogeneity. Explicit references to the existence and the causes of such heterogeneity can be found, for example, in Daniere (1994) -where higher valuations of housing locations near the place of employment, regular source of potable water and public transportation are estimated for low income families in Cairo and Manila-, and in Blamey (1997) where willingness to pay depends on personal attributes like

goods. In fact, their results are directly applicable to the analysis presented in the following pages, given that they obtain (and experimentally validate) distributions of individual valuations for a specific private product's characteristic with a public externality. Inspired on those results we focus on the properties of such distributions to address the issue of (in)compatibility of private and social incentives to modify the consumer's attitude towards the environment.

Economists have warned policy makers on the possible anti-competitive effects of imposing minimum environmental quality requirements due to possible restrictions in the entry of new competitors following higher fixed cost and barriers related with a specific know-how (see Moraga-González and Padrón-Fumero, 2002). However, both the economic literature and actual economic policy-making have ignored the effect of fashion or social trends causing changes in the consumer's environmental consciousness on the intensity of competition among manufacturers and the resulting economic and environmental market efficiency and social welfare. In this chapter, we address this issue. One of our arguments follows standard results on product differentiation, according to which, the heterogeneity of consumer's WTP for a more ecological product will have non-negligible effects on the intensity of competition among manufacturers. Therefore, it is interesting to study exogenous changes in consumer heterogeneity and thus, the economic and environmental efficiencies of markets for private goods.

We adopt a vertical[3] product differentiation framework. We focus on (exogenous) changes in the consumer's WTP for the ecological performance of a product. In particular, the characteristic under consideration is, objectively speaking, contained in one of the two products more than is in the other, and consumers are heterogeneous in the sense that they have different reservation prices for an extra (objective) unit of the characteristic. We focus on three types of increases in consumers' WTP: First, those increasing the consumer's WTP in a uniform way; second, those increasing consumer heterogeneity; third, changes decreasing consumer heterogeneity. Although it would be interesting to study the questions addressed here in a "multi-characteristics space" framework, the only other characteristic we consider is price, which is endogenously determined as the result of competition à la Bertrand between the sellers of the two varieties.

As Cremer and Thisse (1991) show, vertical differentiation models like the one considered here[4] include, as a special case, horizontal differentiation models.[5] This is, mainly, due to the fact that vertical differentiation models of this type combine objective (common for all consumers) and subjective (consumer-specific) valuations of product qualities. This

awareness of environmental need, consequences of and responsibility for personal action and acceptance of policy iniciatives.

[3] Qualitative and quantitative aspects of informative (Grossman & Shapiro, 1984) and persuasive (von der Fehr & Stevik, 1998; Bloch & Manceau, 1999) advertising have been extensively studied in the framework of horizontal differentiation models.

[4] Proposed by Mussa & Rosen (1978).

[5] Although Gabszewicz & Thisse (1986) argue that vertical and horizontal differentiation models are quite different regarding the necessary conditions for the existence of a stable market outcome. In fact, they show that vertical differentiation models give rise to such a stable outcome more frequently than do horizontal differentiation models.

necessary[6] combination of objective - an exclusive feature of vertical differentiation - and subjective - common in all "non-representative consumer" differentiation models - aspects of product quality evaluation by consumers, makes our analytical results easily applicable to horizontal differentiation models.[7]

An important feature of the vertical differentiation model used here is that "increasing a product's quality" would be synonymous - in qualitative terms - with "increasing equally all consumers' WTP for the product". Therefore, the results obtained in the literature on quality choice[8] are directly applicable to changes in consumers' WTP. Yet, the question of why some privately financed advertising messages seem to be directed to certain types of consumers, has not been addressed so far under vertical differentiation.

THE MODEL

Two firms, *a* and *b*, sell two products which are identical in all other aspects, except for their performance with respect to the natural environment. In fact, the products' environmental qualities are assumed to be, respectively, S_a and S_b , which are engineering (objective) measures of the positive[9] impact that each product has on the natural environment, as compared to a similar basic (hostile to the environment) product. The products' qualities are such that $S_a > S_b > 0$. The population of potential (and actual, given the assumptions of the model) consumers is M. Each consumer i maximises her utility U_i, demanding a maximum of one unit of the product from firm $j \in \{a, b\}$ whose product quality and price satisfy:

$$U_i = max\{R + v_i \cdot S_a - P_a, R + v_i \cdot S_b - P_b, 0\} \tag{1}$$

where $R > 0$ is the utility enjoyed by the consumer from consuming a unit of the product before environmental considerations are taken into account; v_i is a consumer-specific parameter indicating consumer i's marginal WTP for a unit improvement of the environmental quality of a product, which is uniformly distributed along the interval $[m, n]$ ($m \geq 0$, $n > m$) with density d. We introduce the notation $L = n - m$, which is the length of the segment along which consumer tastes are uniformly distributed or, alternatively, the consumer heterogeneity parameter. Note that $d \cdot (n - m) = d \cdot L = M$. The term P_j denotes the price

[6] It is straightforward to see why, if consumer heterogeneity were removed, the theoretical framework would require restrictive (non generic) conditions in order to produce results which sustain both firms.

[7] Nevertheless, our model is essentially a vertical differentiation model. That is, any reference concerning the two "aspects" of the model (vertical and horizontal) should not be understood to imply what is meant in the literature by "interaction between vertical and horizontal product differentiation", like, for example, in Degryse (1996).

[8] See, for example, Motta (1993).

[9] That is, following the terminology in Garella & Lambertini (1999), environmental performance is a good rather than a bad characteristic and could be measured, for example, as the inverse of a product's undesirable environmental externality.

fixed by firm j. We assume that R, S_a and S_b are sufficiently high, so that the possibility of consumers with zero consumption is ruled out.[10]

Then, for a pair of prices (P_a, P_b) which are not too different from each other, there will be a consumer i_0, whose environmental consciousness parameter $v_0 \in [m,n]$, makes her indifferent between buying from the less ecological or from the more ecological manufacturer:

$$v_0 \cdot S_a - P_a = v_0 \cdot S_b - P_b \Rightarrow v_0 = \frac{P_a - P_b}{S_a - S_b} \tag{2}$$

Any consumer with $v<v_0$ ($v>v_0$) will feel that the price difference is too high (low) as compared with the environmental quality difference between the two products and she will buy the less (more) green variety. Without loss of generality, we will use the normalisation $S_b=0$, in which case S_a is an environmental quality improvement with respect to the standard variety. We can use this fact and expression (2) to write the two firms' demands:

$$q_a = (n - v_0) \cdot d = \left[n - \frac{P_a - P_b}{S_a} \right] \cdot d$$

$$q_b = (v_0 - m) \cdot d = \left[\frac{P_a - P_b}{S_a} - m \right] \cdot d \tag{3}$$

which, assuming that prices are net of a common, constant marginal cost, except for c which is an *extra unit production cost* paid by the manufacturer of the more ecological variety, give the corresponding profit functions:

$$\Pi_a = (P_a - c) \cdot \left[n - \frac{P_a - P_b}{S_a} \right] \cdot d$$

$$\Pi_b = P_b \cdot \left[\frac{P_a - P_b}{S_a} - m \right] \cdot d \tag{4}$$

[10] Choi & Shin (1992) consider a model in which the market may not be fully covered by the sales of the two firms. However, they consider a model in which this outcome depends on the location - on the line of real numbers - of the interval along which tastes are distributed (parameter m in our model, which is equivalent to parameter θ in their model). This is because they assume that $L=1$. Thus, their concept of "taste diversity" (depending on θ) is different from our concept of "consumer heterogeneity" (reflected on the length of the segment, L). In a more recent article, Wauthy (1996) uses the concept of consumer heterogeneity in a way that is similar to ours. He shows that both outcomes (complete and incomplete market coverage) can be endogenously obtained under conditions concerning the upper and the lower bounds of the segment.

Firms set prices to maximise individual profits. The corresponding Bertrand-Nash equilibrium will satisfy the standard first order conditions giving rise to the system of reaction functions:

$$\partial\Pi_a / \partial P_a = 0 \Rightarrow P_a = \frac{c + n \cdot S_a + P_b}{2}$$

$$\partial\Pi_b / \partial P_b = 0 \Rightarrow P_b = \frac{P_a - m \cdot S_a}{2} \quad (5)$$

We concentrate on three types of changes in consumers' WTP for a more ecological product:

> **TYPE 1:** changes increasing the environmental consciousness, without affecting consumer heterogeneity,
> **TYPE 2:** changes increasing the consumers' environmental consciousness and, at the same time, consumer heterogeneity and
> **TYPE 3:** changes increasing the consumers' consciousness, *decreasing* at the same time consumer heterogeneity.

As we have argued already, the second and the third type of changes may correspond to fashion or social trends that affect, in different ways consumers with different attitudes towards a product's environmental performance. For example, the second type may correspond to trends that increase more the environmental consciousness of the *greenest* among consumers. Similarly, the third type may correspond to trends that increase the consciousness of those of the consumers who were willing to pay less for a greener product. Of course, the examples offered here are only stylised and special -although fairly plausible- types of consumer-specific effects of any social trend causing the aforementioned changes. Studying a larger variety of consumer-specific effects of changes in WTP or different aggregate phenomena which can be modelled as changes in general (non-uniform) distributions of tastes would require a far more complicated exercise producing qualitatively similar results to those reported here.[11].

The solution of (5) is used to derive equilibrium magnitudes[12], leading us to the following conclusion:

> **Proposition 1:** Firm **a** always benefits from type 1 and type 2 changes. The third type will benefit firm **a** only if: $m \cdot S_a > c$. Firm **b** is harmed by the first and the third types of changes, whereas the second type will benefit firm **b** if: $n \cdot S_a > c$.

Benefits to both firms from increasing consumer heterogeneity (second type) require that the most environmentally conscious consumer in the market considers the difference between

[11] Simulations with a more general family of distribution functions (Beta) are available from the authors.

[12] Analytical details on the derivations of equilibrium magnitudes are provided in the appendix.

the two products sufficiently high to justify the extra production costs paid by the manufacturer of the more green product. This is a rather weak condition as compared to the condition for profitability of the third type of WTP changes for the high quality firm, which requires that the *least* environmentally conscious among the consumers would be willing to pay, for the difference between the products, more than the extra unit cost of producing the green product. In fact, if a better environmental performance is not reflected on higher unit production costs, the former of the two conditions holds always, whereas the latter of the conditions is never satisfied.

In the real world, the actual trends definitely contribute in increasing the consumers' WTP for a more ecological product. As we have argued, these trends may have different effects on consumers with different attitudes towards the product's characteristics. Proposition 1 shows that the effects of changes in WTP on consumer heterogeneity will determine, to a great extent, whether a certain social trend increases (or decreases) firms' profits. According to Proposition 1, under reasonably general conditions, both firms benefit from changes increasing consumer heterogeneity, whereas none of them would gain from changes decreasing consumer heterogeneity, unless a rather strong condition holds, under which the manufacturer of the more ecological between the two products benefits only. On the contrary, firms have opposite interests with respect to changes in WTP which leaves consumer heterogeneity unchanged. As we would have expected, changes in consumers' WTP which uniformly increases all consumers' environmental consciousness, benefits the more ecological firm and harms the less ecological one.

Although model parameters were assumed to be such that both firms survive in equilibrium, the results reported so far are directly applicable to the case in which WTP's changes affect one or the other firm's survival possibilities. In fact, changes of the second type would be the only ones which would benefit the less ecological firm. An increase in the consumer's environmental consciousness is not a sufficient condition for changes in WTP to increase the more ecological firm's profits (unless the first condition in Proposition 1 holds). An extra condition for this to happen would be that changes in WTP increase consumer heterogeneity or leave it unchanged.

In order to discuss welfare and policy implications of our results, we will need to calculate the level of social welfare in equilibrium. This will be the sum of the perceived by consumers positive impacts of the two products, market and environmental (objective loss of utility by outsiders and future generations), *minus* the cost of producing the high quality product (with full market coverage, prices are an internal monetary transfer from consumers to producers and their effect on social welfare is zero). Then,

$$SW = d\cdot\left[\int_{m}^{v_0} R\,dv + \int_{v_0}^{m+L}(R + vS_a)\,dv + (S_a - c)\cdot(m + L - v_0)\right] =$$

$$= d\cdot\left[\frac{S_a}{2}\cdot\left((m+L)^2 - \frac{(P_a - P_b)^2}{S_a^{\,2}}\right) + (S_a - c)\cdot\left(m + L - \frac{P_a - P_b}{S_a}\right) + \right] + M\cdot R \qquad (6)$$

This social welfare function takes into account that the social value of a product's environmental quality goes beyond its market valuation and that this is due to the environmental externality on non consumers, future generations, etc.

The third type of changes[13] in WTP is the only one among those considered here which has an unambiguously positive effect on social welfare. The other two types behave in a qualitatively similar way to each other. That is, increasing the consumer's WTP for the environmental quality of a product is socially desirable, as long as the consumers are, initially willing to pay a sufficiently high price for an extra unit of quality.

Given the assumptions under which we study social welfare ($n>c>m$) and, having the above and Proposition 1 in mind, we show the following result:

Proposition 2:
i) *Firm **a**'s and public interests are (in)compatible under:*

- *Type 1 changes, if* $m>m^*$ ($m<m^*$).
- *Type 2 changes, if* $L>L^*$ ($L<L^*$).

ii) *Firm **b**'s and public interests are (in)compatible under:*

- *Type 1 changes, if* $m<m^*$ ($m>m^*$).
- *Type 2 changes, if* $L\notin[L^*,L^{**}]$ ($L\in[L^*,L^{**}]$),

where $L^{**}=c-m$.[14]
iii) *Both firms' interests are incompatible with social welfare maximisation with respect to type 3 changes.*

Our results are applicable to the case of privately financed advertising. Therefore, an alternative way of interpreting Proposition 2 could be:

- As far as changes in m are concerned (for a given degree L of consumer heterogeneity), any effort undertaken by firm b (firm a) in order to decrease (increase) the consumer's marginal WTP for quality is socially desirable, as long as $m<m^*$ ($m>m^*$).
- Alternatively, regarding changes in L, both firms' interests would be (in)compatible with higher levels of social welfare if $L>L^{**}$ ($L<L^*$), whereas, for $L^*<L<L^{**}$, only the high quality firm's higher profitability is compatible with higher levels of social welfare.
- Finally, although changes in WTP of the third type are - under the assumptions made here - unambiguously socially desirable, we would not expect firms to make any effort in that direction.

Some of our conclusions concern difficult to observe data. However, the effects of changes like the ones considered here on outputs and prices are rather easy to trace on price and output data. We observe that only changes of the second type increase both products'

[13] See analytical details in the appendix.

[14] L^{**} is obtained from the condition of profitability of an increase in L for the low quality firm. It is straightforward to obtain by rearranging the condition $c<(m+L)S_a$ in proposition 1. m^* and L^* are defined in the appendix.

prices. The first type has effects of opposite signs on the two products' equilibrium prices. As we would have expected, changes which uniformly increase all consumers' WTP for quality, will increase the price of the more ecological product and decrease the price of the less ecological one. Changes in consumers' WTP of the third type, increase competition between the two firms by decreasing consumer heterogeneity, then decreasing equilibrium prices.

In all cases, the changes in WTP of consumers will increase the difference between equilibrium prices. This result contradicts the claim that smaller price differences reflect more intense competition among firms, given that, in this framework, a larger difference results from the third type of changes, which increases competition and lowers prices.

The market share of the more ecological producer weakly increases in the presence of any of the three types of changes considered. In the case of "type 1" changes, the positive relation between the price and market share for the more ecological firm explains a fact which is often reported in empirical work on differentiated markets. A combination of lower prices and higher market shares in the case of the third type, would be compatible with the contrary phenomenon, that is, a negative correlation between market shares and prices, although the phenomenon would be reversed, if relative, rather than absolute, prices were taken into account. In the presence of changes of "type 3", it is the increase in the market share of the more ecological firm that may compensate (under the first condition in Proposition 1) the firm's losses due to lower prices and make this type of changes profitable for the firm. Similar correlation between market shares and prices can be found to emerge in equilibrium for the less ecological producer. Therefore, we would expect a far more systematic (positive) correlation between market shares and relative (rather than absolute) prices.

A SURVEY

A population of 960 engineering and economics students answered in an anonymous way the following question:

A product is available in the market in two varieties. The only difference between them is that the ***second*** is environmental-friendly.

Suppose that two types of potential consumers exist. Type "*A*" and Type "*E*". The latter are the only ones who are willing to pay an extra amount of money for purchasing the environmental-friendly variety.

Having in mind recent campaigns undertaken by the state on the Spanish TV in order to increase ecological awareness,

Which type of consumer's willingness to pay a higher price for the second variety is affected to a larger extent?

☐ 1) Both consumer types' willingness to pay a higher price for the second variety is equally affected.

☐ 2) Type ***E*** consumer's willingness to pay a higher price for the second variety is affected more.

☐ 3) Type ***A*** consumer's willingness to pay a higher price for the second variety is affected more.

☐ 4) It differs from campaign to campaign, but according to my experience, in the majority of the cases, among the 3 options above, option

1) ☐☐ 2) 3)

is more likely to be true.

☐ 5) I do not want to answer/I do not know.

Table 1 summarizes the results obtained from the above questionnaire.

Table 1. Percentage of responses obtained from the survey

Option	1	2	3	4			5
				4.1	4.2	4.3	
	2.50%	22.50%	33.13%	41.25%			0.62%
				0.62%	17.50%	23.13%	

The question format allows us to distinguish between respondents who think that ecological awareness-increasing campaigns have an unambiguous impact (options 1-3) on consumers of one or the other type (or both of them) and those who think that the impact depends on the specific campaign (options 4.1.-4.3). Aggregate results show that according to our respondents, type 3 effects (which are, according to our theoretical findings the socially desirable ones) are the most frequent. And this is true for both respondents who choose the unambiguous version of the statement (33.13%, choosing option 3) as well as those choosing the ambiguous version (23.13% choosing option 4.3) of it. This indicates that, indeed, the state may have realized the social gains from campaigns targeting the least ecologically conscious among the consumers. However, a significant percentage of respondents (combining the two versions of the statement, 40% of them) feel that the result has been the opposite. Apart from discrepancies in subjective perceptions of the campaigns, heterogeneity of responses may be due to different degrees of exposure to different channels (giving priority to local vs. national level campaigns, etc.). In any case, our main hypothesis concerning the variety of effects that ecological awareness-building campaigns may have on the economic and ecological performance of markets is supported by the survey data. Furthermore, a more careful selection of messages seems to be needed in order to better focus on consumers with lower degrees of ecological consciousness.

CONCLUSIONS

We have collected data on subjective perceptions of the effects of state campaigns on the consumers' ecological awareness. Our focus is on changes in the consumer's willingness to pay for a more ecological product. A strong and a weak version of the statements were offered as an option to the respondents. The former corresponds to an unambiguous effect of the campaigns and the latter to an ambiguous effect. Responses under the two alternative versions reproduce qualitatively similar patters. According to our results, state campaigns on the Spanish TV have targeted more the least ecologically conscious among the consumers. However, a significant percentage of respondents (40%) feel that the contrary is true.

Using our theoretical framework, we can study the effect of a variation in consumers' environmental consciousness on firms' profits, consumer surplus and overall social welfare. As one would expect, we find that, ceteris paribus, the producer of the more environment-friendly variety always prefers selling to a market whose consumers' environmental consciousness is higher. If consumer heterogeneity is kept constant, the producer of the less (more) environment-friendly variety earns lower (higher) profits with campaigns which increase uniformly the environmental consciousness of all consumers. Interestingly, we find that the manufacturer of the less ecological product may also benefit from selling to a market with more environmentally conscious consumers. Increases in consumers' WTP for a higher environmental performance of marketed products, followed by increased consumer heterogeneity may also (and under less restrictive assumptions) benefit both firms. Therefore, markets in which ecologically aware consumers are willing to pay more for a green product are more likely to be more profitable also for non ecologically-oriented manufacturers. That is, markets may fail to create the right incentives for firms to invest in improving their products' performance with respect to the environment.

Welfare analysis shows that exogenous changes in the WTP of consumers, reducing consumer heterogeneity, is socially desirable, but is the least likely to be favoured by private profit-driven actions.

Although we would not like to exaggerate the possibilities of state intervention in such detailed qualitative aspects of social trends, changes in consumers' information or in their income, it should be clear that any means of providing the consumer with more information on the environmental quality of their products should be challenged against our main finding: in most cases, firm strategies would favour advertising campaigns leading to socially undesirable changes.

As Comanor and Wilson (1979) have already pointed out, measures undertaken by firms aiming at reducing competition among them may increase the consumers' WTP for a more expensive product. This unambiguous result concerning the relation between changes in WTP and market power becomes less easy to apply by policy makers in real-world economic policies, if the effect of those changes on consumer preferences and the resulting utility is taken into account. That is, more information about green products may result in higher prices, but people paying higher prices for those products may be happier because this information makes them like the products they consume more. Then, the comparison of *pre-* and *post*-information prices under the assumption that *pre-* and *post*-information products are essentially the same may lead to misleading conclusions.[15] It would seem that, when products have an externality that can be objectively evaluated in terms of social welfare, the effects of changes in the consumers' WTP become easier to assess. Contrary to this conjecture, we find that factors which may increase the consumer's WTP for a green product, may also lead to lower levels of social welfare. As we have seen in this chapter, this may be due to an excessively high extra cost of producing the most ecological variety.

The robustness of our results with respect to the variations considered here and the numerous possible ways of applying this very simple framework to study other issues related with changing the consumers' attitude towards quality, leave a lot of (unexplored) space for

---

[15] Becker & Murphy (1993) analyse advertising as a product sold together with other conventional products, suggesting that an endogenous tastes framework is not necessarily the only way of dealing with changes in consumers' attitudes towards advertised products.

further investigating information changes, social trends, income changes, etc. which may affect consumers' WTP in the vertical product differentiation framework. For the moment, our survey data indicate that there is a lot of space for improving state campaigns in order to intensify and make clearer the message towards the least ecologically conscious among the consumers.

ACKNOWLEDGEMENTS

Financial support by Bancaixa (P1 1B 2001-05), the Generalitat Valenciana (GRUPOS04/13) and the Ministry of Science and Technology (BEC2002-04380-C02-02) is gratefully acknowledged.

REFERENCES

Becker, G., and Murphy, K.M. (1993). A Simple Theory of Advertising as a Good or Bad. *Quarterly Journal of Economics,* 108, 941-964.

Blamey, R. (1997). Contingent Valuation and the Activation of Environmental Norms. *Ecological Economics,* 24, 47-72.

Bloch, F., and Manceau, D. (1999). Persuasive Advertising in Hotelling's Model of Product Differentiation. *International Journal of Industrial Organization,* 17, 557-574.

Camacho, E., García-Gallego, A., Georgantzís, N. and Sabater, G. (2003). An Experimental Test of Response Consistency in Contingent Valuation. *Ecological Economics,* 47, 167-182.

Camacho, E., García-Gallego, A., Georgantzís, N. and Sabater, G. (2004). An Experimental Validation of Hypothetical WTP for a Recyclable Product. *Environmental and Resource Economics*, 27, 313-335.

Choi, C.J., and Shin, H.S. (1992). A Comment on a Model of Vertical Product Differentiation. *Journal of Industrial Economics,* 40, 229-232.

Comanor, W., and Wilson, T.A. (1979). The Effect of Advertising on Competition: A Survey. *Journal of Economic Literature,* 17, 453-476.

Cremer, H., and Thisse, J.F. (1991). Location Models of Horizontal Differentiation: A Special Case of Vertical Differentiation Models. *Journal of Industrial Economics*, 39, 383-390.

Daniere, A. (1994). Estimating Willingness-to-pay for Housing Attributes. An Application to Cairo and Manila. *Regional Science and Urban Economics,* 24, 577-599.

Degryse, H. (1996). On the Interaction between Vertical and Horizontal Product Differentiation. *Journal of Industrial Economics*, 44, 169-186.

von der Fehr, N.H., and Stevik, K. (1998). Persuasive Advertising and Product Differentiation. *Southern Economic Journal,* 65, 113-126.

Gabszewicz, J.J., and Thisse, J.F. (1986). On the Nature of Competition with Differentiated Products. *Economic Journal*, 96, 160-172.

Garella, P., and Lambertini, L. (1999). Good vs. Bad Characteristics in Vertical Product Differentiation. *Economics Letters*, 65, 245-248.

Grossman, G.M., and Shapiro, C. (1984). Informative Advertising with Differentiated Products. *Review of Economic Studies,* 51, 63-81.

Moraga-González, J.L., and Padrón-Fumero, N. (2002). Environmental Policy in a Green Market. *Environmental and Resource Economics,* 22, 419-447.
Motta, M. (1993). Endogenous Quality Choice: Price *vs.* Quantity Competition. *Journal of Industrial Economics*, 41, 113-131.
Mussa, M., and Rosen, S. (1978). Monopoly and Product Quality. *Journal of Economic Theory,* 18, 301-317.
Nyborg, K. (2000). Homo Economicus and Homo Politicus: Interpretation and Aggregation of Environmental Values. *Journal of Economic Behavior and Organization*, 42, 305-322.
Shaked, A., and Sutton, J. (1982). Relaxing Price Competition through Product Differentiation. *Review of Economic Studies*, 49, 3-14.
Wauthy, X. (1996). Quality Choice Models of Vertical Differentiation. *Journal of Industrial Economics*, 44, 345-353.

MATHEMATICAL APPENDIX

Equilibrium Magnitudes

Equilibrium prices are expressed as follows:

$$P_a^e = \frac{1}{3}\cdot\left[2\cdot c + (2n-m)\cdot S_a\right]$$
$$P_b^e = \frac{1}{3}\cdot\left[c - (2m-n)\cdot S_a\right] \tag{A.1}$$

and the corresponding equilibrium outputs:

$$q_a^e = d\cdot\frac{(2n-m)\cdot S_a - c}{3\cdot S_a}$$
$$q_b^e = d\cdot\frac{c-(2m-n)\cdot S_a}{3\cdot S_a} \tag{A.2}$$

Equilibrium outputs will be positive as long as $2n-m > c/S_a > 2m-n$. This condition guarantees that, in equilibrium, there will be a consumer who is indifferent between the two products. That is, $m < v_0^e = \frac{(m+n)}{3} + \frac{c}{3\cdot S_a} < n$.

Then, equilibrium profits are given by:

$$\Pi_a^e = d \cdot \frac{[(2n-m)\cdot S_a - c]^2}{9 \cdot S_a}$$

$$\Pi_b^e = d \cdot \frac{[c-(2m-n)\cdot S_a]^2}{9 \cdot S_a} \qquad \text{(A.3)}$$

We observe that, while equilibrium quantities and profits depend on consumer density (and the total population of the consumers), equilibrium prices and, thus, the environmental consciousness of the *indifferent* consumer depend on relative rather than absolute characteristics of consumer distributions.

Profitability of Changes in Consumers' Heterogeneity

The population density $d=M/(n-m)$ may, also, be expressed in terms of $L=(n-m)$ as $d=M/L$. It will be useful to write equilibrium profits in terms of M, m and L, in order to study changes in WTP affecting equally the WTP of all consumers, which corresponds to an increase in m, keeping M and L constant (first type):

$$\Pi_a^e(m,L) = M \cdot \frac{[(m+2L)\cdot S_a - c]^2}{9 \cdot L \cdot S_a}$$

$$\Pi_b^e(m,L) = M \cdot \frac{[c-(m-L)\cdot S_a]^2}{9 \cdot L \cdot S_a} \qquad \text{(A.4)}$$

The same expressions can be used to study changes in WTP which increase consumer heterogeneity modelled as an increase in L, keeping m and M constant (second type).

Finally, it will be useful to write profits in terms of m, n and M in order to study changes in WTP which decrease consumer heterogeneity modelled as an increase in m, keeping n and M constant (third type):

$$\Pi_a^e(m,n) = M \cdot \frac{[(2n-m)\cdot S_a - c]^2}{9 \cdot (n-m) \cdot S_a}$$

$$\Pi_b^e(m,n) = M \cdot \frac{[c-(2m-n)\cdot S_a]^2}{9 \cdot (n-m) \cdot S_a} \qquad \text{(A.5)}$$

Then, the derivatives of the expressions in (A.4) with respect to m, first, and then with respect to L would give us the effect of the first two types of changes on firms' profits for a given consumer population M:

$$\partial\Pi_a^e / \partial L = -M \cdot \frac{[(m+2L)\cdot S_a - c]\cdot[(m-2L)\cdot S_a - c]}{9\cdot S_a \cdot L^2}$$

$$\partial\Pi_b^e / \partial L = -M \cdot \frac{[(m+L)\cdot S_a - c]\cdot[(m-L)\cdot S_a - c]}{9\cdot S_a \cdot L^2}$$

$$\partial\Pi_a^e / \partial m = 2M \cdot \frac{[(m+2L)\cdot S_a - c]}{9L}$$

$$\partial\Pi_b^e / \partial m = 2M \cdot \frac{[(m-L)\cdot S_a - c]}{9L} \qquad \text{(A.6)}$$

Similarly, the derivatives of the expressions in (A.5) with respect to m will give us the effect of the third type of changes on firms' profits, for a given M:

$$\partial\Pi_a^e / \partial m = M \cdot \frac{[c - m\cdot S_a]\cdot[c-(2n-m)\cdot S_a]}{9\cdot S_a \cdot (n-m)^2}$$

$$\partial\Pi_b^e / \partial m = M \cdot \frac{[c-(3n-2m)\cdot S_a]\cdot[c-(2m-n)\cdot S_a]}{9\cdot S_a \cdot (n-m)^2} \qquad \text{(A.7)}$$

Social Welfare Analysis

Provided that the result obtained from our welfare analysis aims at illustrating a possible deviation between private and social benefits from increasing WTP, we simplify the analysis by introducing a further assumption: we will use the normalisation $S_a=1$. Then, the condition for survival of both firms in equilibrium becomes: $m<c<n$, which implies a plausible situation in which the most environmentally conscious consumer believes that the extra cost of producing the ecological product is justified, whereas the least environmentally conscious among all consumers believes the contrary. Treating the term $M{\cdot}R$ as a constant, we can show that:

$$\frac{\partial SW}{\partial m} = \frac{M}{9L}\cdot\left(7L+3-5(c-m)\right) \qquad \text{(A.8)}$$

$$\frac{\partial SW}{\partial L} = \frac{M}{18L^2}\cdot\left(8L^2+6(c-m)-5(c-m)^2\right) \qquad \text{(A.9)}$$

and that second order conditions are:

$$\frac{\partial^2 SW}{\partial m^2} = \frac{5M}{9L} > 0 \qquad \text{(A.10)}$$

$$\frac{\partial^2 SW}{\partial L^2} = \frac{M}{9L^3} \cdot (c-m) \cdot \left(5(c-m)-6\right) > 0 \tag{A.11}$$

provided that $5 \cdot (c\text{-}m) > 6$.

The roots of the equations $\frac{\partial SW}{\partial m} = 0, \frac{\partial SW}{\partial L} = 0$ are, respectively, $m^* = \frac{5c - 7L - 3}{5}$ which is positive as long as $5c > 7L + 3$, and $L^* = \sqrt{\frac{(c-m)[5(c-m)-6]}{8}}$ which corresponds to a real number as long as the content of the square root is positive (guaranteed, among other, by the conditions: $c > m$ and $[5(c-m)] > 6$).

The expressions in (A.10) and (A.11) indicate that m^* and L^*, may correspond to local *minima* of equilibrium social welfare. Therefore, assuming that m^* and L^* satisfy the conditions for survival of both firms in equilibrium, we can conclude that, for $m>m^*$ ($m<m^*$) and $L>L^*$ ($L<L^*$), equilibrium social welfare is an increasing (decreasing) function of m and L respectively. Furthermore, some straightforward algebra can show that, if m^* and L^* fall out of the intervals satisfying the conditions for survival of both firms in equilibrium, social welfare is an increasing function of both parameters (m and L).

Following the methodology adopted in the previous section, we should, also, express social welfare in terms of parameters m and n. Given the limited insights which can be drawn from the resulting expressions, rather than re-writing SW in terms of m and n, we will focus on the derivatives:

$$\frac{\partial SW}{\partial m} = \frac{M \cdot \Omega}{18(n-m)^2} \tag{A.12}$$

in which, for simplicity, we substituted $\Omega = 6n^2 - 6c + 5c^2 + m^2 + 6n - 10cn - 2mn$, and

$$\frac{\partial^2 SW}{\partial m^2} = \frac{M(n-c)(6-5(n-c))}{9(n-m)^3} > 0 \tag{A.13}$$

if $5 \cdot (n\text{-}c) < 6$. This condition will be compatible with the necessary condition for inequality (A.11) to hold, as long as $c > (n\text{-}m)/2$, which implies that heterogeneity is relatively low compared to the extra unit production costs to be justified. However, expressions (A.12) and (A.13) imply that SW is a strictly increasing function of m, taking positive values for all non negative values of the parameter (a more detailed proof is available from the authors).

In: New Trends in Ecology Research
Editor: A. R. Burk, pp. 135-158

ISBN: 1-59454-379-8

Chapter 6

MATRIX MODELS: A TOOL FOR LANDSCAPE MANAGEMENT?

Lars Westerberg ***[*]*** ***and Uno Wennergren***
IFM/Theory and modelling,
Linköpings Universitet,
SE-581 83 Linköping

ABSTRACT

Matrix models, i.e. expressing systems of linear equations in matrices and using matrix algebra, have been successful in describing the demographic dynamics of a population. The same method can be used to study the effect of space on population growth and spatial distribution. Here, a temporally and spatially discrete matrix model is used to ask general and specific question in landscape ecology. One such question is how habitat amount and arrangement affects population distribution and persistence, for example how habitat loss in various habitat arrangements (random, clumped or fragmented landscapes) affects a population.

We shortly review how matrix models have been used to determine the effect of spatial structure of a population and how this approach have been generalised to a multi-patch system. We demonstrate that a matrix model (including population redistribution, survival and growth) can serve as a general framework for spatial redistribution in a landscape. The results are comparable with existing results from studies of critical thresholds. Existing analytical tools are implemented to determine the sensitivity of e.g. population growth and stable population distribution (eigen-value and -vector sensitivity) to changes in transitions between landscape cells. Some examples and illustrations of how sensitivity values can be used in landscape management are provided.

---

[*] Corresponding author

INTRODUCTION: MATRIX MODELS AND LANDSCAPE ECOLOGY

Matrix models[1] were introduced in population ecology some 60 years ago by H. Bernardelli (1941), E.G. Lewis (1942) and P.H. Leslie (1945, 1948). The number of ecological studies that adopted matrix models was low at start but has since the 1970s grown and is today widely used (Caswell 2001 and references therein). The most famous of the three pioneers is Leslie who studied age-structured populations and showed that the "rate of natural increase r" is closely related to the dominant (real and positive) latent root of the determinant (the eigenvalue) of the Leslie matrix, and that the stable age structure as well as stable birth-rate (real and positive right and left eigenvectors) can easily be calculated (Leslie 1945, 1948). Later, L.P. Lefkovitch (1965) extended this view by generalising the time discretisation process to stable stage-structured populations, indicating that a population can be sub-divided in many ways. Today, the application of population matrix model theory and its recent development is dominated by, and has been thoroughly covered in, the reference work by H. Caswell (2001). It also covers numerous examples of when matrices have been applied in more complex ecological settings than pure demographic problems. In fact, by defining population processes so that they can be expressed in a primitive, irreducible, nonnegative matrix, there will always exist a dominant positive and real eigenvalue (population growth, λ) with corresponding positive and real right (stable age- or stage structure) and left (reproductive value) eigenvectors, according to the Perron-Frobenius theorem (Bapat and Raghavan 1997; Caswell 2001, p. 83). In this chapter we will include movements to study population processes on a spatial scale and exemplify how eigen-values and -vectors can be used to answer questions in landscape ecology.

Matrix models have been applied in spatial ecology[2] by spatially sub-dividing populations into a patches or regions. For multi-regional models see Rogers (1966) and Lebreton (1996), and for metapopulation models Caswell (2001, p. 62). Multi-regional models were developed by A. Rogers and F.J. Willekens (Rogers 1966, 1973, 1974; Willekens 1977; Rogers and Willekens 1978), who studied the influence of demography and net migration on human populations. Their work on stable population structure, reproductive value, and net production was restated in an ecological context and further developed by J.D. Lebreton (1996). Later, Lebreton and colleagues used this matrix framework and a sensitivity analysis to study evolutionary stable dispersal strategies (Lebreton and others 2000) and optimal population control strategies (Brooks and Lebreton, 2001). The multi-regional approached have been extended to acknowledge that demographic and migration events may occur on different time-scales, and in a series of articles R. Bravo de la Parra and colleagues used an aggregation method for discrete models (Bravo de la Parra and others 1995, 1997; Sánchez and others 1995; Sanz and Bravo de la Parra 1998, 1999). Thus, matrix models have provided a framework for asking and answering question in spatial ecology, and have modelled and analysed the resulting spatial structure of a population by assuming discrete time and space.

---

[1] To express systems of linear equations using matrices, matrix algebra and analytical results such as eigenvalues and eigenvectors.

[2] The study of population processes, such as dispersal and resource use that lead to spatial patterns.

There are also examples of matrix models in landscape ecology[3]. A popular approach to study the dynamics of sub-divided populations is dominated by the metapopulation approach (e.g. Hanski and Gilpin 1997) that focuses on extinction and colonisation events and where the sub-populations inhabit suitable habitat patches that are surrounded by inhospitable non-habitat. Spatially explicit population models, but not necessarily using matrix models and theory (e.g. Dunning and others 1995), have to some extent challenged this view by coupling population processes with a landscape, or resource map, and allowing individuals to move and exist at every point in the landscape. Both approaches have been criticised. For example, metapopulation models are patch-oriented and treat the inhospitable habitat as being completely featureless and hostile; yet it may harbour and sustain individuals (Haila 2002; Erotons and others 2003) as well as affect individual dispersal and patch isolation (Wiens 1997; Ricketts 2001). Furthermore, spatially explicit population models suffer from difficulties in parameter estimation, data requirements and possibly magnification of estimation error (Wennergren and others 1995; Kareiva and others 1997; but see Wiegand and others 2004). However, both approaches deal with movements between spatial positions and demographic processes within each position and, even though metapopulation models are usually formulated as continuous extinction/colonisation processes, both approaches may be handled by using discrete matrix models (Lebreton and Gonzales-Davila 1993; Hanski and Ovaskainen 2000; Caswell 2001; Westerberg and Wennergren 2003).

Recent development of the "metapopulation capacity" concept have applied a matrix framework to a spatially realistic Levins metapopulation model (Hanski and Ovaskainen 2000, 2003; Ovaskainen and Hanski 2001, 2003; Ovaskainen 2003). In their work, colonisation and extinction events, patch size and arrangement are parameters used to formulate a "landscape matrix". The eigenvalue of the landscape matrix is the capacity of a certain landscape to maintain a viable population, i.e. the metapopulation capacity. Among many other features, this model allows a straightforward calculation (using the eigenvalue and the right eigenvector) of each patch's importance to the threshold condition for metapopulation persistence and analytical sensitivity analyses. Other metapopulation and multi-patch studies have used sensitivity (and elasticity) analysis for stage- and patch-structured matrices (Caswell 2001, p. 206) for example to: assess the ecotoxicological consequences on population growth, age- and spatial population structure in a river network (Chaumot and others 2003); model two-patch Peregrine falcon population viability under different management regimes (Wootton and Bell 1992); analyse the sensitivity of a two patch sea scallop system to predation, dispersal and growth (Barbeau and Caswell 1999); identify probable causes to deviations from observed spatial structure in a plankton transport model (Lewis 1997, in Caswell 2001, p. 68). Thus, matrix models have spurred much research and provided tools for studies of spatial and landscape ecology.

The development of multi-regional and metapopulation models, and the possibilities from using a matrix framework, has inspired us to continue to study the capacity of these tools. In this chapter we exemplify how a matrix model of habitat specific demographics and population redistribution can be used as a starting point to study general and applied questions in landscape ecology. One general landscape ecological question is if there are general rules regarding the effect of habitat amount and arrangement on population viability (e.g. Fahrig 2002; Flather and Bevers 2002). Or, from another perspective: are there different

[3] The effect of (large) spatial pattern on population processes.

responses to habitat amount and arrangement for different life histories, for example for different reproductive schedules (in this chapter when movement-, growth- and survival-events occur on different time scales)? Applied questions that focus on a specific landscape could ask whether total population growth will change as a consequence of changes in that landscape. A closely related applied question is to what extent will a landscape change cause the population to redistribute within the landscape and if a larger fraction of the population will end up somewhere else in the landscape? Here, we give some examples of how to use matrix theory to answer general and applied questions, and we show the potential of eigenvalue and eigenvector sensitivity analysis (Tanner and others 1994; Wennergren and others 1994; Caswell 2001, p. 247) when applied in landscape ecology.

METHOD: THE LANDSCAPE MATRIX MODEL

A general matrix model for population redistribution in a heterogeneous landscape that is constant over time (Westerberg and Wennergren 2003; Westerberg and others, to appear), is:

$$\mathbf{N}_{t+1} = \mathbf{M}\mathbf{N}_t \quad (1.)$$

$\mathbf{N}_t$ is the population vector at time t=0, 1, 2 ... . The transition[4] matrix **M** contains the proportion moving between landscape cells (m_{ij}, $i \neq j$) and the probability to remain in cell j (m_{jj}). A matrix that is non-negative (all matrix elements $m \geq 0$), irreducible (all cells are directly or indirectly reciprocally connected), and primitive (the greatest common divisor of transition loop lengths between cells is one) have got one dominant, real and positive eigenvalue ($\lambda_1 > |\lambda_j|$, $j>1$) according to the Perron-Frobenius theorem (Bapat and Raghavan 1997; Caswell 2001, p. 83). In equation (1), **M** is non-negative, irreducible and primitive. By applying periodic boundary conditions, all columns of **M** sum to one, which leads to a dominating eigenvalue (λ_1) of one, i.e. the population is just redistributed without losses or growth.

The lattice landscape, or resource map, **H** consists of $x \times y = k$ grid cells that are either of preferred or secondary habitat quality (the ratio of preferred habitat is denoted p and the amount is $p \times k$). The landscape cells are indexed by their order (i) in a column vector after stacking the columns of **H**, i.e. vec(**H**). All matrices and vectors are arranged accordingly and thus the dimension of **N** and all other vectors is k^2. **M** and all other matrices are of dimension $k^2 \times k^2$.

Below we discuss four different cases: firstly when reproduction, survival and movement act on the same time scale (Case 1); secondly when survival and movement act on a much faster time scale than reproduction (Case 2). The two sub-cases are with mortality (Case 1b and 2b) and without (Case 1a and 2a). Examples of organisms with "slow" reproduction separated from "fast" movement and survival could be predatory ground beetles in agricultural ecosystems with a discrete reproduction event each season. Organisms with simultaneous reproduction, movements and survival could be agricultural pests such as

[4] In the remainder of this chapter, the terms *transition* and *movement* will differ. The latter stress the behavioural process of changing position, while the former also accounts for demographic events and represent the probability to, or proportion that, change position.

aphids. To study these cases, the original model (eq. 1) is modified by including population growth (**B**) and survival (**S**). Assume that the dynamics of a population can be expressed by:

$$\mathbf{N}_{t+1} = (\mathbf{S} \circ \mathbf{M})^{\mathrm{T}} \mathbf{B}(\mathrm{T}) \mathbf{N}_t \qquad (2.)$$

Individuals first reproduce instantaneously, then survive and disperse during the T "fast" projections of one "slow" projection interval t. The order of **B** and (**S**∘**M**) in eq. (2) affects asymptotic population distribution but not growth. $\mathbf{B}(\mathrm{T})=\mathrm{diag}(b_1, b_2, \ldots b_k)$ is the diagonal reproduction matrix, where $b_i=[1,+\infty]$ is the reproduction of the subpopulation in cell i before the T "fast" projection periods. Since $b_i \geq 1$, **B** only express population growth; $b_i=1$ represents either no reproduction (Case 1, overlapping generations without mortality within one season) or no net reproduction (Case 2, one non-overlapping generation per season). It is also important how **B**(T) scale with T (see below). The survival matrix, **S**, describe the survival rate for remaining in cell j (s_{ij}), and when moving from cell j to i (s_{ij}, $i \neq j$). The Hadamard product (∘) denotes element-by-element multiplication of the **S** and **M** matrices. When reproduction, survival and movement acts on the same time scale (Case 1), then T=t=1, while T>>1 when survival and movement is much faster than reproduction (t=T, 2T, ...; Case 2).

When all demographic events take place on the same time scale (T=1; Case 1), we use $(\mathbf{S} \circ \mathbf{M})^1 \mathbf{B}(1)=\mathbf{A}$ to rewrite eq. (2) into $\mathbf{N}_t=\mathbf{A}^t\mathbf{N}_0$. The elements of **A** are the product of the movement probability (m_{ij}) from j to i, survival probability (s_{ij}) during that move, and reproduction in start cell j (b_j): $a_{ij}=b_j \times s_{ij} \times m_{ij}$. The dynamics of $\mathbf{N}_t=\mathbf{A}^t\mathbf{N}_0$ can be expressed using all eigenvalues and left (**v**) or right (**w**) eigenvectors of **A**:

$$\mathbf{A}^t \mathbf{N}_0 = \sum_i C_i \times \left(\lambda_i^{\mathbf{A}}\right)^t \times \mathbf{w}_i^{\mathbf{A}} = \sum_i C_i \times \mathbf{v}_i^{\mathbf{A}} \times \left(\lambda_i^{\mathbf{A}}\right)^t,$$

where C_i is a constant depending on $\mathbf{N}_0$. The long-term dynamics will be dominated by the largest eigenvalue and its corresponding eigenvectors:

$$\mathbf{N}_t = \mathbf{A}^t \mathbf{N}_0 \approx \left(\lambda_1^{\mathbf{A}}\right)^t \times \mathbf{w}_1^{\mathbf{A}} = \mathbf{v}_1^{\mathbf{A}} \times \left(\lambda_1^{\mathbf{A}}\right)^t. \qquad (3.)$$

The population growth is $\lambda_1^{\mathbf{A}}$, and the population will persist if $\lambda_1^{\mathbf{A}} \geq 1$ (i.e. always persist for Case 1a) as long as enough individuals survive while moving to cells where $b \geq 1$ (Case 1b). Both growing and declining populations will, at equilibrium, be distributed according to $\mathbf{w}_1^{\mathbf{A}}$ (the stable spatial distribution). A similar, but more complex approach is outlined in Lebreton and others (2000), including density dependent growth and with the purpose to study evolutionary stable dispersal strategies. They studied an asymptotic population behaviour evaluated at carrying capacity where population growth is one, which is a different equilibrium than the one derived from eq. (3). We will use eq. (3) because of its simplicity.

If survival and movements acts on a much faster time scale (T>>1; Case 2), we may base the full model on $(\mathbf{S} \circ \mathbf{M})^{\mathrm{T}} \mathbf{B}(\mathrm{T})=\mathbf{Q}$. The dominating eigen-value and -vector of **Q** describe its asymptotic behaviour (λ_1^{Q} is population growth and $\mathrm{w}_1^{\mathrm{Q}}$ is the population distribution) during any period of length T, thus:

$$\mathbf{N}_t = \mathbf{Q}^t\mathbf{N}_0 \approx \left(\lambda_1^{\mathbf{Q}}\right)^t \times \mathbf{w}_1^{\mathbf{Q}}.$$

For large values of T, during a projection starting with $\mathbf{B}(\mathrm{T})\mathbf{N}_0$, we expect the asymptotic behaviour of $\mathbf{S}\circ\mathbf{M}=\mathbf{L}$ to have the greatest influence on population distribution. If T is large enough, we temporarily drop $\mathbf{B}(\mathrm{T})$ and approximate the population dynamics using only $\mathbf{L}$'s dominating eigen-value and vector, $\mathbf{L}^{\mathrm{T}}\mathbf{N}_0 \approx \left(\lambda_1^{\mathbf{L}}\right)^{\mathrm{T}} \times \mathbf{w}_1^{\mathbf{L}}$. Now, knowing the fast dynamics, we assume that $\mathbf{B}(\mathrm{T})$ operates on the stable population distribution, i.e. $\mathbf{N}_0 \simeq \mathbf{w}_1^{\mathbf{L}}$. Hence, assuming stable distribution, we get the approximation:

$$\mathrm{N}_{\mathrm{T}} = (\mathrm{S}\circ\mathrm{M})^{\mathrm{T}}\mathrm{B}(\mathrm{T})\mathrm{N}_0 \approx \mathbf{B}(\mathrm{T})\left(\left(\lambda_1^{\mathbf{L}}\right)^{\mathrm{T}} \times \mathbf{w}_1^{\mathbf{L}}\right), \tag{4.}$$

Equation (4) is not a completely satisfactory model, since it can only project $\mathbf{N}_t$ for t=T, but not for t>T. A better formula would allow the population to redistribute between reproduction events, e.g.: $\mathbf{N}_{t+\mathrm{T}} = \mathbf{B}(\mathrm{T})\left(\left(\lambda_1^{\mathbf{L}}\right)^{\mathrm{T}} \times \left(\mathbf{w}_1^{\mathbf{L}} \times \sum_i \mathbf{N}_{i,t}\right)\right)$, t=0, T, 2T, ... A more formal treatment of time and space discrete matrix models with processes acting on different time scales is presented in e.g. Sanz and Bravo de la Parra (1999). By using an aggregation method, they show that such a complex system may be reduced as long as one of the two processes, that works on different time scales, are conservative (i.e. do not change the total population number). We do not see any reasons why their approach would not work if none of the two processes are conservative, but it is beyond the scope of this chapter. Even though a model based on the aggregation method could better approximate $\mathbf{Q}$, eq. (4) is transparent and has got some features that satisfy a comparison with eq. (1) and (3).

In absence of mortality ($\mathbf{S}$=1; Case 2a) $\lambda_1^{\mathbf{L}}=1$, and if the right eigenvector is scaled to sum to one, then population growth (R) can be determined by:

$$R = \sum_i \mathbf{B}(\mathrm{T}) \times \mathbf{w}_{1,i}^{\mathbf{L}}.$$

If mortality is included (Model 2b), then $\lambda_1^{\mathbf{L}}<1$ and the population will persist only when:

$$R = \left(\lambda_1^{\mathbf{L}}\right)^{\mathrm{T}} \times \sum_i \mathbf{B}(\mathrm{T}) \times \mathbf{w}_{1,i}^{\mathbf{L}} \geq 1.$$

Thus, eq. (4) provides an approximation to growth ($R\approx\lambda_1^{\mathbf{Q}}$, for large T) and to the stable distribution at the end of the season or right before reproduction ($\mathbf{w}_1^{\mathbf{L}} \approx \mathbf{w}_1^{\mathbf{Q}}$, for large T) both when survival is included (Case 2b) and when not (Case 2a). The stable population distribution from eq. (4) may be a bad predictor of asymptotic population structure during other parts of the projection interval T.

To have the same population growth in Case 1 and 2 after for example a season, irrespective of the value of T (and consequently different **B**(T)), it is important how **B**(T) scale with T. Since $(\mathbf{BL})^d \neq \mathbf{B}^d\mathbf{L}^d$, thus $\mathbf{B}(\mathrm{T})^{1/\mathrm{T}} \neq \mathbf{B}(1)$. An approximation for not too small values of T is to find a **B***(T) to seasonal growth **B**(*d*T) such that:

$$\sum_i \mathbf{B}(d\mathrm{T}) \times \mathbf{w}_{1,i}^{\mathbf{L}} = \left(\sum_i \mathbf{B}^*(\mathrm{T}) \times \mathbf{w}_{1,i}^{\mathbf{A}}\right)^d,$$

where $\mathbf{w}_1^{\mathbf{A}}$ and $\mathbf{w}_1^{\mathbf{L}}$ are defined above.

Simulations

We simulate transitions (i.e. the product of reproduction, survival and movements) in 16×16 lattice landscapes, created by the software RULE (that use the mid-point displacement algorithm; Gardner 1999), and where habitat is arranged either random or fractal. The landscapes have a ratio of p=0.1, 0.5 or 0.9 preferred habitat cells and are either random or with a Hurst dimension of 0 or 1 (fragmented and clumped landscapes respectively). Ten replicates of each landscape setup resulted in 90 simulated landscapes.

The models were implemented using Matlab (MathWorks 2001). Movements in landscapes, for all four cases, were approximated using a simple diffusion equation:

$$m_{ij} = \left(1/4\pi D_{\mathbf{H}i}\right) \times e^{\left(-r_{ij}^2/4D_{\mathbf{H}i}\right)},$$

where r_{ij} = [0, $r_{\max}$] is the distance between cell j and i and D is the habitat dependent diffusion constant: D=0.1 if $\mathbf{H}_i$ is preferred habitat, else D=1. No movements beyond $r_{\max}$ =7 are possible, i.e. the diffusion function is truncated at that distance. The survival, for Case 1b and 2b, is a simple linear function of distance to target cell:

$$s_{ij} = 1 - r_{ij}/2r_{\max},$$

where only moving individuals are at risk (since s_{ii}=1). Finally, seasonal reproduction for preferred and secondary habitat cells was set to:

b_i =1.25 where $\mathbf{H}_i$ is preferred, else b_i = 1.

Instantaneous population growth takes place after T=1 or T=20 for Case 1 and 2 respectively, and season length is 20 "fast" projection intervals. For Case 1, eq. (3) reads:

$$\mathbf{N}_{20} = \left(\lambda_1^{\mathbf{A}}\right)^{20} \times \mathbf{w}_1^{\mathbf{A}}, \text{ for } \mathbf{A} = (\mathbf{S} \circ \mathbf{M})\mathbf{B}(1),$$

and Case 2 eq. (4) is:

$$\mathbf{N}_{20} = \mathbf{B}(20)\left(\left(\lambda_1^{\mathbf{L}}\right)^{20} \times \mathbf{w}_1^{\mathbf{L}}\right), \text{ for } \mathbf{L} = (\mathbf{S} \circ \mathbf{M}).$$

We chose to adjust reproduction for Case 1, $\mathbf{B}^*(1)$, so that overall population growth at the end of the season is the same as in Case 2 $\mathbf{B}(20)$:

$$\sum_i \mathbf{B}(20) \times \mathbf{w}_{1,i}^{\mathbf{L}} = \left(\sum_i \mathbf{B}^*(1) \times \mathbf{w}_{1,i}^{\mathbf{A}}\right)^{20}.$$

$\mathbf{B}^*(1)$ had to be adjusted for each unique landscape by either solving the polynomial or by numerically finding the values of $\mathbf{B}^*(1)$ that satisfy the equality. We chose the latter method.

RESULTS: POPULATION RESPONSE TO LANDSCAPE, MORTALITY TND TIME SCALES

Here we report results (from the four cases Case 1a, b and 2a, b) on seasonal population growth or persistence ($\lambda_1^{\mathbf{A}}$, $\lambda_1^{\mathbf{L}}$ or R), population distribution ($\mathbf{w}_1^{\mathbf{A}}$ or $\mathbf{w}_1^{\mathbf{L}}$) to study general rules regarding the effect of habitat amount and arrangement on population viability. We also exemplify how analytical sensitivity and elasticity analysis of the transition matrix (Caswell 2001, p. 206; see separate section below) for specific landscapes can be used to determine how population growth and distribution change with altered transitions probabilities.

Growth, Stable Distribution and Resilience

Seasonal population growth (Figure 1) was naturally higher in landscapes with more preferred cells. Population growth decreased 35-65% when mortality was included, even when preferred habitat was abundant. The latter might seem to be an unexpectedly high decrease since 80% of the sub-population in preferred cells (where the population aggregated, Figure 2) did not move and were thus not exposed to mortality. However, during a season more than 99% of the population was redistributed and the combined effect of mortality on growth is the product and not sum of survival events.

There was also an effect of habitat arrangement (random, clumped and fragmented landscapes): population growth was lower in randomly configured landscapes and especially at low ratio of preferred habitat. Highly clumped landscapes (Hurst dimension of 1) enjoyed the highest population growth because a larger proportion of the population occupied preferred cells (Figure 2). A lower proportion of the population in preferred cells in landscapes with a low ratio of preferred cells (and when preferred cells were scattered) explain why values for repeated reproduction (the b's making up $\mathbf{B}(1)$; Figure 1) had to be higher to achieve the same seasonal population growth as when T=20. The probable mechanism behind is that dispersal from preferred cells was mostly to other preferred cells in landscapes with abundant or clumped preferred cells.

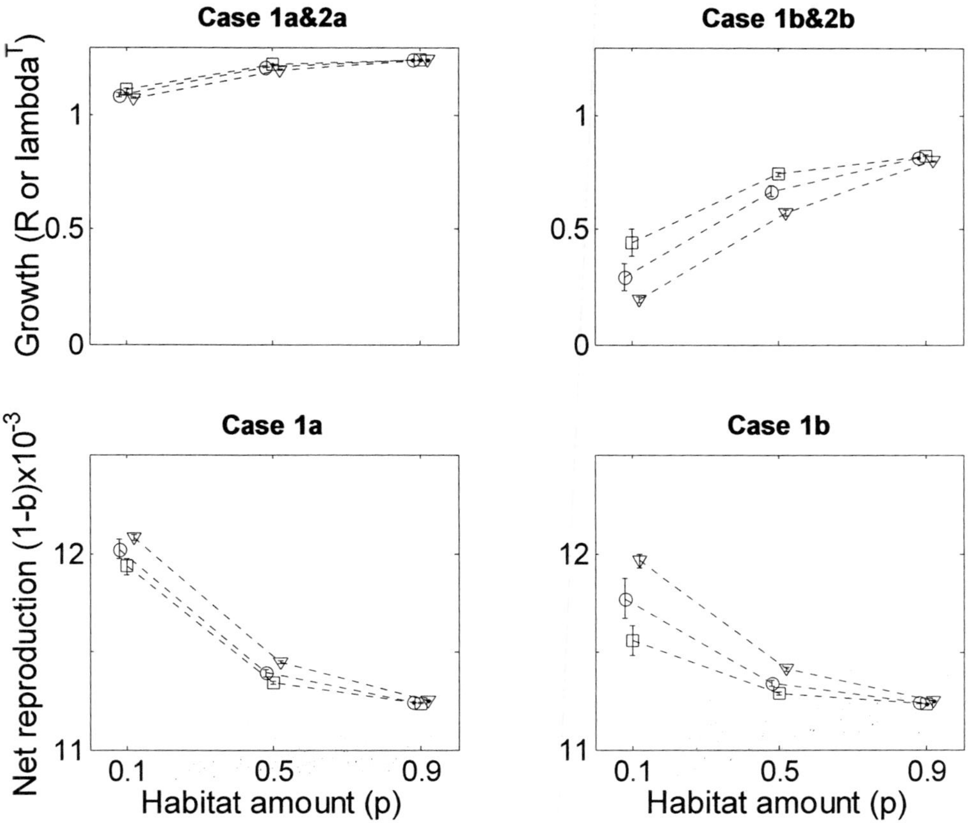

Figure 1.Population growth per season (T=20) and net reproduction value in preferred cells (1-*b*) when reproduction occur every time step (Case 1). The growth value for Case 1 simulations was the dominating eigenvalue for projection interval t=1, to the power of season length (T=20 projection intervals). Growth value (*R*) for Case 2 simulations were calculated using $R = \left(\lambda_1^{\mathrm{L}}\right)^{\mathrm{T}} \times \sum_i \mathbf{B}(\mathrm{T}) \times \mathbf{w}_{1,i}^{\mathrm{L}} \geq 1$. Reproduction values for Case 1 were scaled so that seasonal growth was similar for Case 1a&2a, and 1b&2b respectively (observed difference of < 1% between Case 1 $(\lambda_1)^{20}$ and Case 2 *R*). See text for details on scaling reproduction values. Amount habitat is the ratio of preferred to secondary habitat in the landscapes. Circles, squares, and triangles are the average results (±standard deviation) in landscapes with Hurst dimension 0 and 1, and in random landscapes respectively.

Populations with simultaneous reproduction, survival and movement (Case 1) generally needed higher reproduction values ($\mathbf{B}(1) > \mathbf{B}(20)^{1/20}$) in preferred cells (Figure 1). This indicated that the distribution of populations with "fast" reproduction led to less effective landscape utilisation, compared to populations with "slow" reproduction. The latter had equal seasonal growth as the former, but with lower reproduction value. Another measure of how well the population use available resources in the landscape, is the proportion in preferred landscape cells ($\sum_g \mathbf{w}_{1,g}$, where *g* are all positions *i* where $\mathbf{H}_i$ is a preferred habitat cell). The ratio of the population in preferred cells (Figure 2) is strongly associated with population growth (Figure 1) since growth directly depends on proportion of population in preferred cells and indirectly on the amount and aggregation of preferred cells. Population distributions for

Case 1 and 2 are calculated differently ((**S**○**M**)**B** =**A** which gives $\lambda_1^{\mathbf{A}}$ and $\mathbf{w}_1^{\mathbf{A}}$, and **S**○**M**=**L** which gives $\lambda_1^{\mathbf{L}}$ and $\mathbf{w}_1^{\mathbf{L}}$), but proportion in preferred habitat differ only by 3%. The minute difference is a consequence of the relatively small values of the **B** matrix, which makes it close to a unit matrix (Case 1). When mortality was included, the population concentrated in preferred landscape cells because dispersal (and consequently also mortality) was higher from secondary cells.

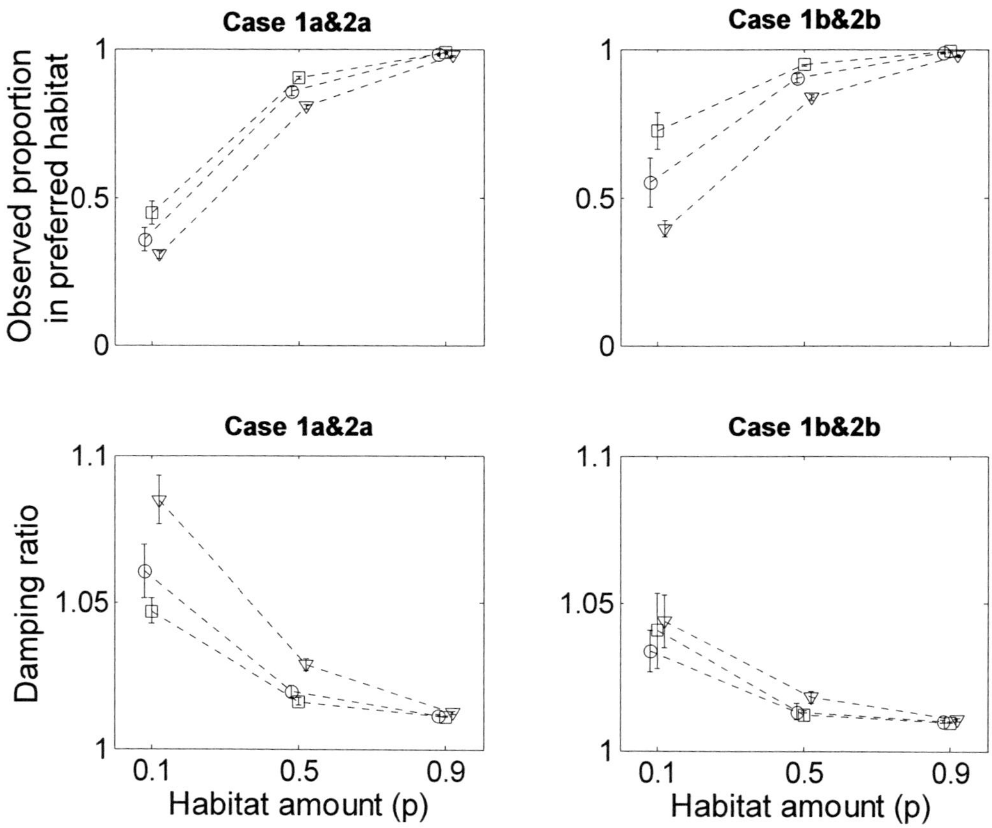

Figure 2. The ratio of the population that occupied preferred cells was given by the right eigenvector (w) that corresponded to the dominating eigenvalue. The damping ratio ($\lambda_1/|\lambda_2|$) measured how fast the population dynamics approached its asymptotic state, described by the dominating eigen-value and -vector. A high value indicated a large difference between the two largest eigenvalues and that the eigenvector was a reasonable approximation. Values close to one were the result from eigenvalues of similar sizes, thus slowly approaching asymptotic behaviour. Values from Case 1a&2a, and 1b&2b were similar (differed less then 3% and 0.1% for observed proportion in preferred cells and damping ratio) because reproduction values were scaled to give the same seasonal growth (see Figure 1). Amount habitat is the ratio of preferred to secondary habitat in the landscapes. Circles, squares, and triangles are average results (±standard deviation) in landscapes with Hurst dimension 0 and 1, and in random landscapes respectively.

The damping ratio (Caswell 2001, p. 95) between the two largest eigenvalues, $\rho = \lambda_1 / |\lambda_2|$, is a measure of how fast the system approaches asymptotic behaviour. A high value means that the leading eigenvalue has got a large influence, and that the population distribution soon approaches the dominating eigenvector. Another interpretation of the damping ratio is resilience, or how fast the population returns to the asymptotic state after a disturbance (Westerberg and others to appear). Landscapes with a high proportion of the population moving will reach asymptotic state sooner than when a low proportion is moving. For example: landscapes with few preferred cells that retain individuals (the proportion staying in secondary cells is much lower than proportion staying in preferred) have got a high damping ratio, opposite to when preferred habitat is abundant (Figure 2). The damping ratio is higher in random landscapes because transitions from preferred cells is more often to secondary cells in random than clumped landscapes. Finally, time to reach the asymptotic state increase when mortality is introduced, because a part of the moving population disappear which reduces the redistribution speed. The general message is that matrix models of populations in landscapes characterised by long transient time (i.e. when preferred cells are abundant and clumped) might be less reliable in applied situations.

Eigenvalue Sensitivity and Elasticity

The analytical sensitivity of the eigenvalues to similar sized perturbations in matrix elements can be determined analytically by (Caswell 2001, p. 206):

$$\partial\lambda_1/\partial a_{ij} = \left(v_{1,i} \times w_{1,j}\right)/\langle \mathbf{w}_1, \mathbf{v}_1 \rangle$$

where **v** and **w** are the corresponding left and right eigenvector respectively, and $\langle \mathbf{w}_1, \mathbf{v}_1 \rangle$ is the scalar product of $\mathbf{w}_1$ and the complex conjugate of elements in the $\mathbf{v}_1$-vector. Sensitivity analyses identify the effect from different factors on model output by estimating the local slope of λ_1 as a function of each matrix element. The sensitivities will always be equal to or larger than zero since the matrix elements are defined so that increased matrix values have got positive effect on population growth. These analytical sensitivities do not necessarily hold for larger perturbations, but are reliable to assess the effect of a small perturbation of a transition to λ_1. Elasticity values are scaled sensitivities that measure the effect of perturbations proportional to the matrix element. The elasticity of the eigenvalue is calculated by $\left(a_{ij}/\lambda_1\right) \times \left(\partial\lambda_1/\partial a_{ij}\right)$, and elasticity values can be regarded as contributions to the eigenvalue (Caswell 2001, p. 229). We chose to report elasticity for Case 1 only, since Case 2 eigenvalues does not contain reproduction. Further, elasticity values are generally more intuitive than sensitivities in this setting, because it is easier to think of proportional rather than absolute perturbations of transitions. Though, both analyses gave the same qualitative results. Sub-sets of elasticities and sensitivities were added to get a general picture.

The sum of eigenvalue elasticity for transitions within and out of preferred cells was high when preferred habitat was abundant, but decreased as preferred habitat decreased (Figure 3). For Case 1a, a slight general increase in transition within and from preferred cells, when preferred habitat was scarce, had a smaller effect than a similar increase of transitions within and from secondary cells; maybe because the large number of secondary cells had a higher influence on the eigenvalue. This was not the case when mortality was added (Case 1b); probably because a higher proportion of the population occupy preferred cells (cf. Figure 2 and 3). The same argument can be used to explain the slight differences between clumped, fragmented and random landscapes: as preferred cells got more aggregated, the proportion of population in preferred cells increased. Thus, secondary cells are important when abundant, either because of high elasticity to eigenvalue (Case 1a) or because they trigger movements that lead to mortality and lowered population growth (Case 1b).

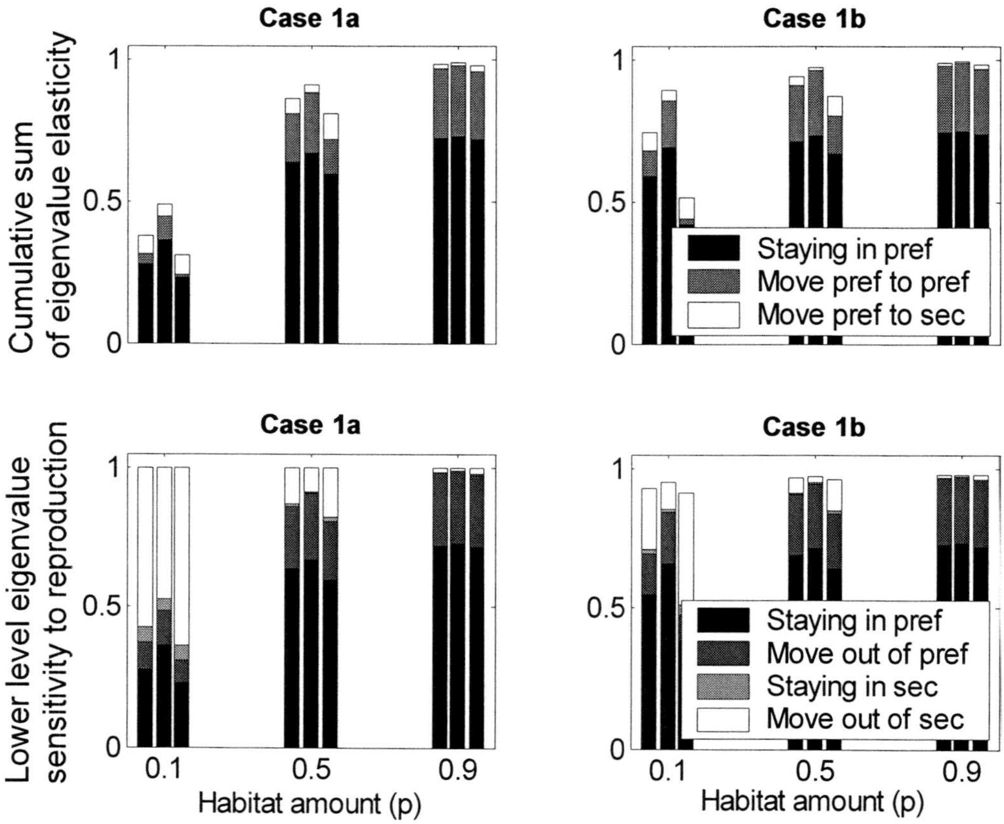

Figure 3. The cumulative sum of eigenvalue elasticity to perturbations in matrix elements that corresponded to staying in preferred cells (black), moving from preferred to other preferred cells (gray), and moving from preferred to secondary landscape cells (white). Total elasticity sum to one, thus the bars show contribution to population growth (λ_1). Elasticity for Case 2 was excluded since reproduction was omitted from the approximation formula (eq. 4). The lower level sensitivity of the eigenvalue to perturbations in reproduction values in matrix elements corresponding to staying in preferred cells (black), moving from preferred to other preferred cells (dark grey), staying in secondary cells (light grey) and moving from preferred to secondary landscape cells (white). The bars are in all plots grouped so that landscape with habitat arrangement of Hurst dimension 0 and 1 is first and second respectively, and random landscapes are last. The groups are arranged in order of the amount of preferred to secondary habitat, i.e. the ratio of preferred habitat in the landscapes.

Matrices **A** (eq. 3) and **L** (eq. 4) contain compounded elements (reproduction-, survival- and movement-values) that do not necessarily correspond to an ecological relevant parameter, which can make interpretations of sensitivity and elasticity analysis difficult. Sometimes a lower level sensitivity analysis (Caswell 2001, p. 218) can be useful:

$$d\lambda_1/dx = \sum_{k,l}\left(\left(\partial\lambda_1/\partial a_{k,l}\right)\times\left(\partial a_{k,l}/\partial x\right)\right).$$

Since $a_{k,l}=f(b, s, m, ...)$, x can be elements from matrix **B**, **S** and **M**, but only for element positions (ij) where e.g. b-elements have the same value. The resulting sensitivities of the eigenvalue to reproduction values (Figure 3) show almost identical results as the eigenvalue elasticity. This supports the conclusion stated above that the population density of a cell is important to eigenvalue sensitivity since it determine overall population growth. To increase population growth in landscapes with little preferred habitat, either improve reproduction in secondary cells (no mortality, Case 1a) or reproduction in preferred cells (with mortality, Case 1b).

Eigenvector Sensitivity

Eigenvector sensitivity (Wennergren and others 1994; Caswell 2001, p. 247) is a measure similar to eigenvalue sensitivity and is calculated (after scaling the eigenvectors so that the scalar products $<\mathbf{w}_i,\mathbf{v}_i>=1$ and $<\mathbf{w}_i,\mathbf{v}_j>=0$, $i\neq j$) by:

$$\partial\mathbf{w}_1/\partial a_{ij} = \mathbf{w}_{1,j}\times\sum_{k\neq 1}\left(\mathbf{v}_{k,i}\times\left(\lambda_1-\lambda_k\right)^{-1}\times\left(\mathbf{w}_j-\mathbf{w}_1\times\sum_l \mathbf{w}_{j,l}\right)\right) \quad (5.)$$

All eigen-values and -vectors are used. The resulting eigenvector sensitivity is a $k^2\times k^2\times k^2$ dimension tensor containing the sensitivity of each population densities in cell l ($\mathbf{w}_{1,l}$) to perturbation of elements in a_{ij} that describe transitions from cell j to i (i.e. $\partial\mathbf{w}_{1,l}/\partial a_{ij}$). The derivatives can be both negative and positive since changes of the transition matrix may both decrease and increase the relative population density in a cell. Both eigen-value and -vector sensitivities should have been adjusted for Case 2a, since matrix columns sum to one (Markov chain matrix; Caswell 2001, p. 251). However, such adjustments were cumbersome to perform for eigenvector sensitivities and did not change the general picture of the results for eigenvalue sensitivities, thus we neglected to perform such adjustments.

A general measure of the sensitivity of densities in preferred and secondary habitat to changes in transitions was the mean of the sensitivity of preferred and secondary $\mathbf{w}_1$-element to any transitions (Figure 4). That is, the average sensitivity of $\mathbf{w}_{1,g}$ (where g are all positions i where $\mathbf{H}_i$ is a preferred habitat cell) to any transitions: $\left(p\times k\right)^{-1}\times\sum_{l=g}\left(k^{-2}\times\left(\sum_i\sum_j\left(\partial\mathbf{w}_{1,l}/\partial a_{ij}\right)\right)\right)$, and similarly for secondary cells. The results (Figure 4) show that at low amount of preferred cells, the densities in preferred habitat were more (negatively) sensitive to a general change of transition values than was population density in secondary habitat. The sign of sensitivities imply that a general increase in transitions (matrix element values) would decrease the density in preferred cells (cf Figure 5, cell specific eigenvector sensitivity). Further, densities in preferred cells in highly clumped landscapes seemed to be more sensitive to changes than other arrangements, and the sensitivity of densities in preferred cells decrease with increasing amount of preferred habitat. Finally, caution should be taken so that comparisons between eigenvector sensitivities are made between matrices with dominating eigenvalues of the same magnitude (i.e. $0<\lambda_1<1$, $\lambda_1=1$ or $\lambda_1>1$; Wennergren and others 1994).

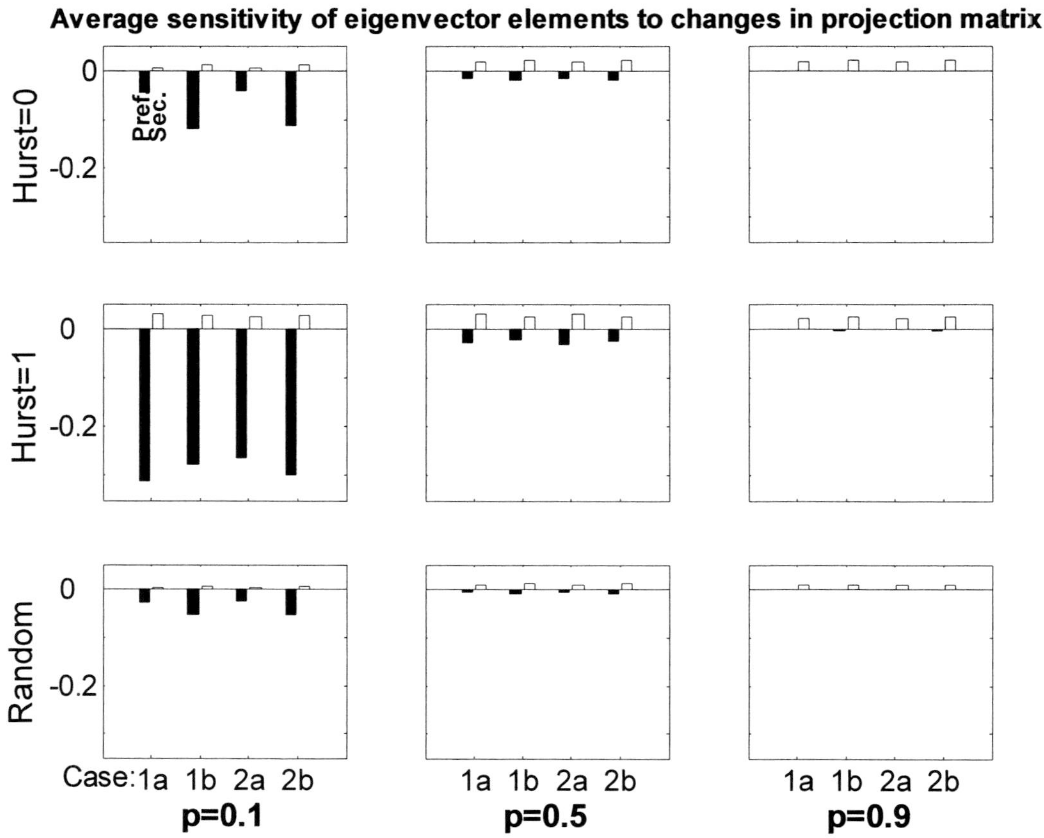

Figure 4. The average sensitivity to perturbation of the projection matrix of eigenvector elements of preferred (filled bars) and secondary (white bars) habitat. Sensitivities in landscapes of different habitat arrangement (Hurst dimension of 0 or 1, or random configuration) and habitat amount (*p* is the ratio of preferred cells) are given in rows and columns, respectively. Sensitivity values for different cases are given in each plot. Negative sensitivities imply that a general positive change of the transition values of the projection matrix leads to a general decreased population density in that group of cells.

Three Examples of Sensitivity and Elasticity Analyses of Specific Landscapes

We chose to further exemplify the use of eigen-value and -vector sensitivities using three examples (Figures 5-7).

In Figure 5, eigenvalue elasticity and sensitivity is reported together with eigenvector sensitivity. The effect on population growth (eigenvalue sensitivity and elasticity) of changed transition values to and from each landscape cell shows a larger effect from increasing transitions to and from clumped preferred cells. Thus, increasing reproduction, survival or import by movement from other cells (export movements must sum to one) naturally improved population growth, and especially if the increase was made in preferred habitat.

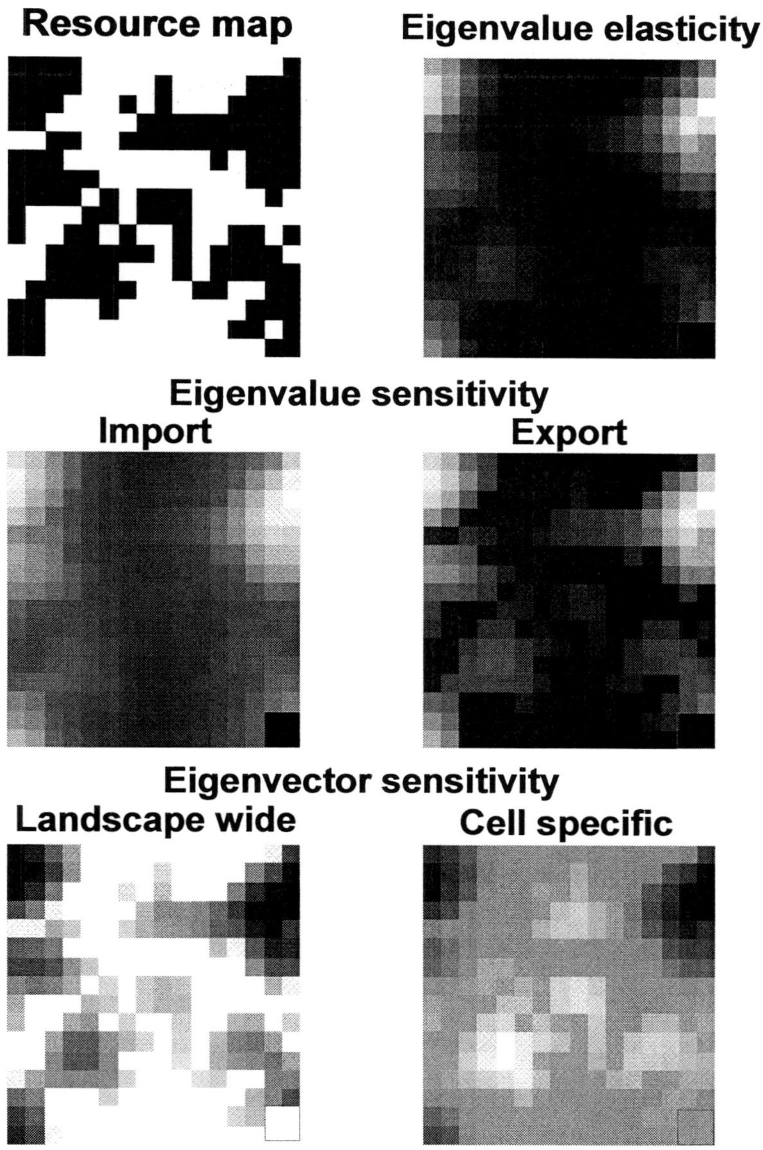

Figure 5. The sensitivity of population growth (eigenvalue) and population distribution (eigenvector) to changes in transitions for a population complying with Case 1b assumptions (repeated reproduction events during the season and with mortality when dispersing). The resource map is the physical layout of preferred habitat (black) in the landscape (habitat amount p=0.5, and with Hurst dimension=0). The brightness of cells indicate sensitivity and elasticity value relative to "zero sensitivity/elasticity" in the encircled four lowest cells to the left in each landscape: brighter shades indicate positive and darker shades negative values. For example, brighter cells in the eigenvalue elasticity plot indicate that changes in transitions to/from that cell will have a large effect on population growth. Darker cells for landscape wide eigenvector sensitivity indicate a general negative sensitivity of the population distribution when transitions to/from that cell are altered. Brighter cells for cell specific eigenvector sensitivity show where the population density will increase as a response to a general increase of transition values.

Landscape wide eigenvector sensitivity (Figure 5) is the general change in population distribution when transitions to and from one cell is changed (by adding import $\sum_i \sum_j \left(\partial \mathbf{w}_{1,i} / \partial a_{lj} \right)$ and export $\sum_i \sum_j \left(\partial \mathbf{w}_{1,i} / \partial a_{jl} \right)$ for each cell l=1, 2 ...,k^2). Changing transitions to and from clumped preferred cells is negative and of greatest magnitude, while population distribution is generally insensitive to altering transitions to and from secondary habitat. The cell specific eigenvector sensitivity (Figure 5) is the magnitude and direction of population density change of a cell as a result of general change in transition rate ($\sum_i \sum_j \left(\partial \mathbf{w}_{1,l} / \partial a_{ij} \right)$ for each cell l=1, 2 ...,k^2). A positive general change in transition rate would decrease the population density in clumped preferred cells, but increase density in less clumped cells. A probable reason is that the population can not effectively aggregate in clumped preferred cells when transition rates increase, but export some individuals to less clumped preferred cells.

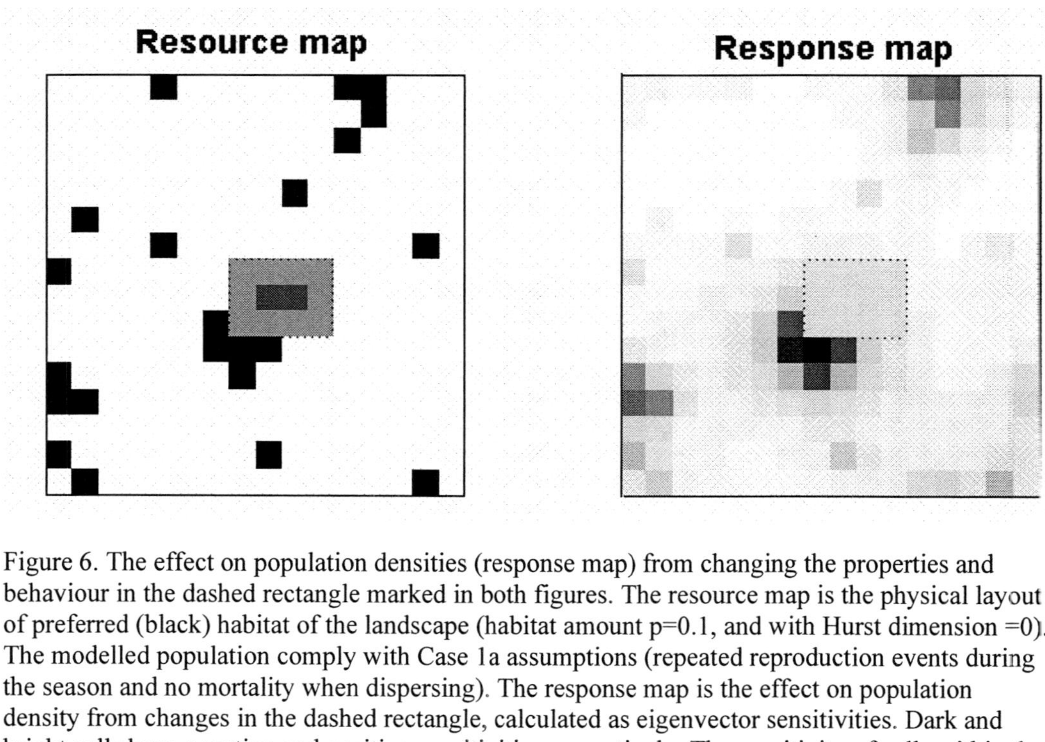

Figure 6. The effect on population densities (response map) from changing the properties and behaviour in the dashed rectangle marked in both figures. The resource map is the physical layout of preferred (black) habitat of the landscape (habitat amount p=0.1, and with Hurst dimension =0). The modelled population comply with Case 1a assumptions (repeated reproduction events during the season and no mortality when dispersing). The response map is the effect on population density from changes in the dashed rectangle, calculated as eigenvector sensitivities. Dark and bright cells have negative and positive sensitivities respectively. The sensitivity of cells within the dashed rectangle is not calculated and its shade represents zero sensitivity. For example, if export from the dashed rectangle is negatively affected, population densities will increase in cells darker than cells in the dashed rectangle.

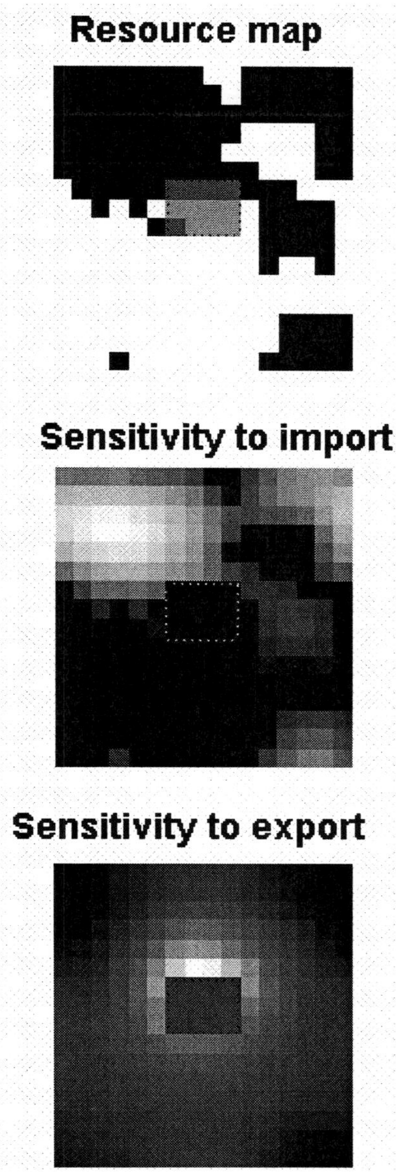

Figure 7. The effect on population densities (response map) within the dashed rectangle, from changes in transitions into the rectangle, calculated as eigenvector sensitivities. The landscape consists of 50% preferred habitat (p=0.5) with a Hurst dimension of 1. The population inhabiting the landscape was modelled using Case 2b (one reproduction event per season and with mortality when dispersing). The brightness of the area inside the dashed rectangle corresponds to zero sensitivity. Darker and brighter cells have negative and positive sensitivities respectively. For example: increase import from bright cells increase population density in target cells; increase export to brighter cells, or decrease export to the darkest cells to increase population density in target cells.

Secondly, lets say that the interaction of specific area with the surrounding landscape is affected, as it would be e.g. from clear-cutting or fencing, how will the stable population densities in the landscape change (Figure 6)? The sensitivity of each cell to changes in the affected area were calculated by summing the columns denoting sensitivity to export from affected cells ($\sum_{j=g}\sum_{i}\left(\partial \mathbf{w}_{1,l}/\partial a_{ij}\right)$, for each cell l=1, 2 ...,k^2, $l \neq g$ where g denote the indices of affected cells). The response map in our example show that clumped preferred areas at quite some distance might be positively affected by a decreased export from the affected area. The mechanism behind is unclear but it could be that the contribution to population density from the affected area is larger for secondary than preferred cells. Thus, the relative proportion of the population increase in clumped preferred cells.

Lastly, imagine that the population density of an area has to change. What transitions should be targeted? Figure 7 shows that the population density in the target area is reduced most if import from bright cells (from preferred habitat where a great deal of the population reproduction takes place) is reduced. Increasing export to neighbouring preferred habitat would increase density of target area, probably because individuals are likely to remain close and return. These sensitivities were calculated by adding eigenvector sensitivities for the targeted cells and plot the sum of rows and columns (import $\sum_{i=g}\partial \mathbf{w}_{1,i}/\partial a_{ij}$, and export $\sum_{i=g}\partial \mathbf{w}_{1,i}/\partial a_{ji}$, for each cell j=1, 2 ...,k^2, $j \neq g$ where g denote the indices of cells in the target area) of the resulting matrix.

Discussion

By adding space to ecology, landscape ecology has invoked new demands and perspectives on models and theory. It is not merely a question of a new dimension of complexity, but also a question of what measures and analytical tools can be useful and relevant. This chapter show that when matrix models and sensitivity analysis are transferred from classical demographic models into spatial models, new perspectives and applications arise.

We have studied how a matrix model can serve as a general framework for spatial redistribution in a landscape under certain assumptions. One assumption is that the habitat specific movement behaviour is maintained during the fast projection interval T. The organism cannot detect habitat boundaries during that period, instead it change behaviour as a response to the resource level it experiences after a time step of length (T). Another assumption is that net reproduction occurs in preferred habitat only and that mortality depends on exposure during movements only. Effect of slight violations of these assumptions can be estimated by sensitivity or elasticity analysis, and these suggest that the assumption on reproduction is important (Figure 3). In general, violations should not affect results much, for example when making conclusions on persistence (i.e. between generation dynamics). Fahse and others (1998) claim that slow population processes (reproduction) is slaved by fast processes (movements), thus population growth ultimately depends on the stable structures that emerge from movements. Our results showed similar results since the population growth seemed to depend on population density in preferred cells (cf Figure 1 and 2). Finally, the approximation formula for Case 2 (eq. 4) was satisfactory. Population distribution emerging from seasonal redistribution without growth was very similar to population distribution accounting for population growth because growth was so small relative to the spatial redistribution. Using the full model for Case 2 would have been equally good, but in this way, we had fewer parameters to assess and got more transparent results.

When may matrix models work? The damping ratio indicated that matrix models converged quicker to asymptotic behaviour the faster the population is moving, i.e. when habitat amount is low or when habitat is arranged randomly. The quicker the population converges, the better the eigen-values and -vectors describe the system. The matrix model can still accurately describe population dynamics, even if damping ratio is low, and population distribution can be achieved by numerical simulation. Then the matrix-vector multiplication of the recurrence formula is a way to model population dynamics without using analytical tools.

Landscape Effect on Population Distribution

Both habitat amount and arrangement affected population distribution and growth, but to different degrees. Population growth (λ_1 and R), adjusted reproduction value ($\mathbf{B}(1)$ for Case 1) and population ratio in preferred cells for various habitat arrangement (random, clumped or fragmented landscapes, Figure 1 and 2) were usually more affected by habitat amount changes than by changes in landscape arrangement. But at low amount of habitat, the arrangement of preferred cells became important.

These results are consistent with some results from other landscape models. For example, Flather and Bevers (2002) reported that habitat arrangement became more important for population size below persistence threshold, where habitat arrangement was even more important than habitat amount when using more detailed descriptions of arrangement. They used a population-based model and simulated between-season dynamics with competition for space and with dispersal mortality. That model and our matrix model differ in several aspects (between year dynamics, density dependence and population growth by numerical simulations), but are to a large degree similar, which is the reason why the two models arrive at the same general conclusions regarding the relative importance of habitat amount and arrangement. Metapopulation models that use colonisation and extinction processes produce different results. In a review by Fahrig (2002), colonisation-extinction models were found to need a much higher increase of habitat in fragmented landscapes to escape extinction (i.e. habitat arrangement is important). The reason is that colonisation events decreased with habitat loss in colonisation-extinction models. In a matrix model, habitat loss implies increased redistribution of the population that leads to higher effect from dispersal mortality and, as a consequence, lowered population growth.

The eigenvalue sensitivity is a powerful tool in classical demographic population ecology and has been widely applied. When applied in landscape ecology it remains useful, for example to find landscape areas or transitions that are important for population growth. The importance of preferred and secondary habitat types can be estimated and reveal what landscape changes may be most effective. Eigenvalue sensitivity and elasticity has previously been applied to patch systems (e.g. Wootton and Bell 1992; Barbeau and Caswell 1999; Chaumot and others 2003) but not on a landscape level before, and figure 3 and 5 show its potential at landscape level with different types of habitats. Population growth elasticity (and lower level sensitivity of reproduction) to transitions within and from preferred cells dominated when dispersal mortality was included. This indicates that increased transitions, and especially reproduction, within or between neighbouring preferred cells would increase population growth (Figure 3). Also, the difference in magnitude between population growth elasticities (and lower level sensitivity of reproduction) for random and clumped landscape arrangements indicate that a larger change of matrix elements must take place in random landscapes to have the same impact as a change in clumped landscapes (Figure 3). A final conclusion drawn from eigenvalue elasticities is that secondary cells are important in landscapes: either through high sensitivity to reproduction in secondary cells (without mortality, Case 1a) or by increasing redistribution and reducing population growth (with mortality, Case 1b). Thus, depending on the landscape and movement patterns of the specific organism, we may expect different impact from amount and arrangement of secondary habitat. This can also imply that differences in the life history between organisms (e.g. "slow" or "fast" reproduction and dispersal mortality) can explain differences in how they respond to changes in the landscape.

The Use of Matrix Models and Eigenvector Sensitivities

Usually, the population redistribution between age-classes (the eigenvalue and eigenvector sensitivity) is mostly interesting to studies of fisheries and other harvesting problems. We have already mentioned the few ecological papers (Tanner and others 1994; Wennergren and others 1994; Caswell 2001, p. 247) that actually have applied the eigenvector sensitivity in classical demographic models (it is more common in field of mechanics). Yet in landscape ecology it is a different situation, where population distribution in space, and the proportion of the population found in a specific habitat or patch, is an important and valuable response variable. The sensitivity of this population distribution to transition changes (local movement, mortality and reproduction), can be used in many different ways, and we have showed some fairly straightforward and easily applied examples to typical applied problems. The effect on population densities from changes of a specific area within the landscape (Figure 6) may be compared to disturbing the landscape by for example a road, a building or a clear-cut. Figure 7 depicts a different angle of the same problem: the need for more or less of the population in a specific area. This situation may correspond to optimizing the distribution of natural enemies to fields with pest species in organic farming.

Eigen-vector and -value sensitivity (see above) can be combined to achieve further insight in the landscape dynamics (Figure 5). In most cases we expect these two measures to correlate fairly well on landscape level. We expect that areas with a large impact on population growth may also have large impact on the distribution of the population. Hence, areas where the mismatch between eigen-value and -vector sensitivity is greatest are the most interesting. In these areas, it is possible to change population distribution over the landscape without changing the growth rate, or vice versa. As a result, the existence of such areas would make empirical studies of the effect from fragmentation on population growth difficult. The sensitivity of resilience, or damping ratio, can be contrasted with population growth or distribution to evaluate any tradeoffs between increasing population growth and changing population distribution. It is likely that changes in landscape or transitions (e.g. adding or rearranging habitat) decrease the damping ratio, which may lead to a situation where asymptotic behaviour never is reached.

In our example we have two different types of life histories. One where reproduction and dispersal occurs at the same timescale, and one with long periods of dispersing and utilizing landscape resources between the reproductive periods. This difference makes the interplay between landscape features and life history different, but since reproduction was scaled to give equal seasonal growth we can only speculate on what could happen. An organism with simultaneous dispersal and reproduction should have high correlation between dispersal and reproduction relative habitat type. When reproduction is "slower" than movements, trade offs between dispersal and reproduction are possible since the organism may use the secondary habitats between its reproductive periods. In the matrix formulation of different timescales one may follow this by studying how the eigenvector change along the set of matrix used for a complete period. The distribution of the population over the landscape will oscillate during such a period (compare with Tuljarpurkar 1985).

CONCLUSIONS

Can matrix models become a tool for landscape managers? The answer is yes and maybe no. Yes, because it provides transparent and general answers to quite complicated questions spanning over several scales. It can also guide the efforts to estimate population process parameters (see Figures 5-7). No, since we can hardly expect any organism to comply with a matrix model: individuals of real populations in real landscape surely exhibit a more complex behaviour. However, the dimensions of a real and notoriously complex system must be reduced for us to understand it, and a matrix model of demographic and spatial population processes is a good way to start. And, if we in the future can show that a real and notoriously complex system actually can be approximated with a reduced system, we may state: yes it's a perfect tool.

ACKNOWLEDGEMENTS

We thank the Swedish Foundation for Strategic Research (SSF, NGSSC) and the Swedish Council for Forestry and Agricultural Research (SJFR) for financial support.

REFERENCES

Barbeau, M. & Caswell, H. (1999). A matrix model for short-term dynamics of seeded populations of sea scallops. *Ecological Applications*, 9, 266-287.

Bernardelli, H. (1941). Population waves. Journal of Burma Research Society, 31, 1-18.

Bapat, R. B. & Raghavan, T. E. S. (1997). *Nonnegative matrices and applications. Encyclopedia of Mathematics and its Applications.* New York, NY: Cambridge University Press.

Bravo de la Parra, R., Auger, P. & Sánchez, E. (1995). Aggregation methods in discrete models. *Journal of Biological Systems*, 3, 603-612.

Bravo de la Parra, R., Sánchez, E. & Auger, P. (1997). Time scales in density dependent discrete models. *Journal of Biological Systems*, 5, 111-129.

Brooks, E. N. & Lebreton, J. D. (2001). Optimizing removals to control a metapopulation: application to the yellow legged herring gull (Larus cachinnans). *Ecological Modelling*, 136, 269-284.
Brotons, L., Mönkkönen, M. & Martin, J. M. (2003). Are fragments islands? Landscape context and density-area relationships in boreal forest birds. *The American Naturalist*, 162, 343-357.
Caswell, H. (2001). *Matrix population models: construction, analysis, and interpretation* (2nd ed.). Sunderland, MA: Sinauer Associates Inc.
Chaumot, A., Charles, S., Flammarion, P. & Auger, P. (2003). Ecotoxicology and spatial modelling in population dynamics: an illustration with brown trout. *Environmental Toxicology and Chemistry*, 22, 958-969.
Dunning, J. B., Stewart, D. J., Danielson, B. J., Noon, B. R., Root, T. L., Lamberson, R. H. & Stevens, E. E. (1995). Spatially explicit population models: current forms and future uses. *Ecological Applications*, 5, 3-11.
Fahrig, L. (2002). Effect of habitat fragmentation on the extinction threshold: a synthesis. *Ecological Applications*, 12, 364-353.
Fahse, L., Wissel, C. & Grimm, V. (1998). Reconciling classical and individual based approaches in theoretical population biology: a protocol for extracting population parameters from individual-based models. *The American Naturalist*, 152, 838-852.
Flather, C. H. & Bevers, M. (2002). Patchy reaction -diffusion and population abundance: the relative importance of habitat amount and arrangement. *The American Naturalist*, 159, 40-56.
Gardner, R. H. (1999). RULE: map generation and a spatial analysis program. In J. M. Klopatek, & R. H. Gardner (Eds.), *Landscape ecological analysis: Issues and Applications* (pp. 280-303). New York, NY: Springer.
Haila, Y. (2002). A conceptual genealogy of fragmentation research: from island biogeography to landscape ecology. *Ecological Applications*, 12, 321-334.
Hanski, I. A. & Gilpin, M.E. (1997). *Metapopulation Biology: ecology, genetics, and evolution*. San Diego, CA: Academic Press.
Hanski, I. A. & Ovaskainen, O. (2000). The metapopulation capacity of a fragmented landscape. *Nature*, 404, 755-758.
Hanski, I. A. & Ovaskainen, O. (2003). Metapopulation theory for fragmented landscapes. *Theoretical Population Biology*, 64, 119-127.
Kareiva, P, Skelly, D. & Ruckelshaus, M. (1997). Reevaluating the use of models to predict the consequences of habitat loss and fragmentation. In S. T. A. Pickett, R. S. Ostfield, H. Shachak and G. E. Likens (Eds), *The ecological basis of conservation: heterogeneity, ecosystems and biodiversity* (pp.156-166). New York, NY: Chapman & Hall.
Lebreton, J. D. (1996). Demographic models for subdivided populations: the renewal equation approach. *Theoretical population biology*, 49, 291-313.
Lebreton, J. D. & Gonzales-Davila, G. (1993). An introduction to models of subdivided populations. *Journal of Biological Systems*, 1, 389-423.
Lebreton, J. D., Khaladi, M. & Grosbois, V. (2000). An explicit approach to evolutionarily stable dispersal strategies: no cost of dispersal. *Mathematical Biosciences*, 165, 163-176.
Lefkovitch, L. P. (1965). The study of population growth in organisms grouped by stages. *Biometrics*, 21, 1-18.

Leslie, P. H. (1945). On the use of matrices in certain population mathematics. *Biometrica*, 33, 183-212.

Leslie, P. H. (1948). Some further notes on the use of matrices in population mathematics. *Biometrica*, 35, 213-245.

Lewis, C. V. (1997). Biological-physical interactions on Georges Bank: plankton transport and population dynamics of the ocean quahog, Arctica islandica. *Ph.D. Dissertation*, joint program in Oceanography, Massachusetts Institute of Technology and Woods Hole Oceanographic Institution.

Lewis, E. G. (1942). On the generation and growth of a population. *The Indian Journal of Statistics*, 6, 93-96.

MathWorks (2001). MATLAB, version 6.5. *www.mathworks.com*.

Ovaskainen, O. (2003). Habitat destruction, habitat restoration and eigenvector-eigenvalue relations. *Mathematical Biosciences*, 181, 165-176.

Ovaskainen, O. & Hanski, I. (2001). Spatially structured metapopulation models: global and local assessment of metapopulation capacity. *Theoretical Population Biology*, 60, 281-302.

Ovaskainen, O. & Hanski, I. (2003). How much does an individual habitat fragment contribute to metapopulation dynamics and persistence? *Theoretical Population Biology*, 64, 481-495.

Ricketts, T. H. (2001). The matrix matters: effective isolation in fragmented landscapes. *The American Naturalist,* 158, 87-99.

Rogers, A. (1966). The multiregional matrix growth operator and the stable interregional age structure. *Demography*, 3, 537-544.

Rogers, A. (1973). The mathematics of multiregional demographic growth. *Environment and Planning*, 5, 3-29.

Rogers, A. (1974). The multiregional net maternity function and multiregional stable growth. *Demography*, 11, 473-481.

Rogers, A. & Willekens, F. J. (1978). The spatial reproductive value and the spatial momentum of zero population growth. *Environment and Planning*, 10, 503-518.

Sánchez, E., Bravo de la Parra, R. & Auger, P. (1995). Linear discrete models with different time scales. *Acta Biotheoretica*, 43, 465-479.

Sanz, L. & Bravo de la Parra, R. (1998). Variables aggregation in time varying discrete models. *Acta Biotheoretica*, 46, 273-297.

Sanz, L. & Bravo de la Parra, R. (1999). Variables aggregation in a time discrete linear model. *Mathematical Biosciences*, 157, 111-146.

Tanner, J. E., Hughes, T. P. & Connell, J. H. (1994). Species coexistence, keystone species, and succession: a sensitivity analysis. *Ecology*, 75, 2204-2219.

Tuljapurkar, S. D. (1985). Population dynamics in variable environments VI. Cyclic environments. *Theoretical Population Biology*, 28, 1-17.

Wennergren, U., Forsling, P. & Weinerfelt, G. (1994). Comparative sensitivity analysis of stable stage structures and reproductive values. *Bulletin of Mathematical Biology*, 56, 945-957.

Wennergren, U., Ruckelshaus, M. & Kareiva, P. (1995). The promise and limitation of spatial models in conservation biology. *OIKOS*, 74, 349-356.

Westerberg, L., Östman, Ö. & Wennergren, U. Movement effects on equilibrium distribution of a habitat generalist in heterogeneous landscapes. *Ecological Modelling*, to appear.

Westerberg, L. & Wennergren, U. (2003). Predicting the spatial distribution of a population in a heterogeneous landscape. *Ecological Modelling*, 166, 53-65.

Wiegand, T., Revilla, E. & Knauer, F. (2004). Dealing with uncertainty in spatially explicit population models. *Biodiversity and Conservation*, 13, 53-78.

Wiens, J. A. (1997). Metapopulation dynamics and landscape ecology. In I. A. Hanski & M. E. Gilpin (Eds.), *Metapopulation biology: ecology, genetics, and evolution* (pp. 504, 43-62). San Diego, CA: Academic Press.

Willekens, F. J. (1977). Sensitivity analysis in multiregional demographic models. *Environment and Planning*, 9, 653-674.

Wootton, J. T. & Bell, D. A. (1992). A metapopulation model of the peregrine falcon in California: viability and management strategies. *Ecological Applications*, 3, 307-321.

In: New Trends in Ecology Research
Editor: A. R. Burk, pp. 159-182
ISBN: 1-59454-379-8

Chapter 7

DECISION MAKING WITH ENVIRONMENTAL INDICES

Dana L. Hoag,[1,*] Catherine Keske-Handley,[2] James Ascough II[3] and Lynne Koontz[4]

Department of Agricultural and Resource Economics
Colorado State University
Department of Agricultural and Resource Economics
USDA-Agricultural Research Service
Great Plains System Research Unit, Fort Collins, Colorado
U.S. Geological Survey, Biological Resource Division
Policy Analysis and Science Assistance Program, Fort Collins, Colorado

ABSTRACT

Since Ott's seminal book on environmental indices (1978), the use of indices has expanded into several natural resource disciplines, including ecological studies, environmental policymaking, and agricultural economics. Nevertheless, despite their increasing use in natural resource disciplines, researchers and public decision makers continue to express concern about validity of these instruments to capture and communicate multidimensional, and sometimes disparate, characteristics of research data and stakeholder interests.

Our purpose here is to demonstrate how useful indices can be for communicating environmental information to decision makers. We discuss how environmental indices have evolved from simple to multicriteria formats, and then propose two advancements, the impact matrix and disparate stakeholder management. We provide examples of simple and multicriteria indices that have been used by policy decision makers. We then build the Impact Matrix (IM), which can account for many indices at one time. The IM was shaped from the concept of a financial risk payoff matrix and is applied to ecosystem risk here. Finally, we show how indices can be used in the IM and other formats for disparate stakeholder management, where indices of stakeholder preferences are systematically integrated with biologically- and physically-based indices.

---

[*] Corresponing Author: Professor, Dona L. Hoag, Department of Agricultural and Resource Economics, Colorado State University, Fort Collins, Colorado 80523-1172 dhoag@colostate.edu 970-491-5549

AN INTRODUCTION TO ENVIRONNEMENTAL INDICES

One of the earliest and most comprehensive discussions of environmental indices was presented by Ott in 1978 in "Environmental Indices: Theory and Practice." Indices, as Ott points out, can help balance a simplistic presentation of data and results with a comprehensive presentation of data and results. When stakeholders have a desire to understand environmental data, they typically want a specific question answered. For example:

- "Are environmental conditions improving are worsening over time?"
- "Is the environment degraded by industry or government?"
- "Are natural resources enhanced by government programs to protect it?"

Unfortunately, it is not always easy to provide simple answers. Often times a complete assessment of multiple environmental considerations resides in a 100-page report, which is difficult to compress and make easy to understand. Furthermore, it may be a challenge for untrained professionals or laypersons to understand. According to Ott, scientists are frequently hesitant to discuss the results of a comprehensive scientific study in simplistic terms, and even a thorough report can be subjected to a researcher's interpretive biases. Hence, although readers may desire more extensive information to answer their questions, they may have neither the time nor the skills to process elaborate answers.

Although it has been more than 25 years since the influential environmental index handbook by Ott (1978) was published, banter continues about the most appropriate approach to communicate environmental data to various stakeholders and a wide variety of audiences. While the discussion has helped improve indices, environmental problems have become more complex. For example, the spectrum of the audience—and stakeholders—with an interest in environmental data reporting has expanded. Rather than being confined to a small circle that included academics, legislators, and a few constituents with interests left or right of center, there are now a diverse number of stakeholders with an interest in environmental data, including:

- Locally organized organizations such as homeowners groups
- Non-governmental, membership driven organizations, such as The Nature Conservancy, The World Wildlife Federation, and the National Rifle Association.
- Hundreds of local, state, and federal government agencies
- Journalists
- Educators
- International audiences

In addition to the broadening of the environmental audience and stakeholder groups, the widespread use of the Internet has made citizens more educated—and impatient. With a few keystrokes into a good search engine, environmental information is available in a nanosecond, and thousands of documents are readily available with minimal research effort. Thus, stakeholders are able to swiftly review a broad range of information, from peer-reviewed academic research to government publications, editorials and up-to-the-minute news information.

With the tremendous search engine capabilities, audiences are also seeking the most time efficient path to satisfy their hunger for information. Despite the fact that a multiple-page comprehensive report may provide the most expansive amount of information and be readily available, many stakeholders (particularly those who are disconnected from the academic arena) are inclined to review the documents that simultaneously provide both the closest match to their word search AND the most concise, accurate amount of information.

What most people want is a concise, accurate synopsis, so that information can be communicated efficiently, accurately, and in a user-friendly format. An index appears to fulfill these objectives quite nicely. Ott's 1978 definition of an index is:

> "Ideally, an index or an indicator is a means devised to reduce a large quantity of data down to its simplest form, retaining essential meaning for the questions that are being asked of the data. In short, an index is designed to simplify."

Although an index appears to fulfill the "timeless" need for communicating a large quantity of information in a concise format, indices can be controversial. As noted by Hoag et al. (2002), the mere task of trying to simplify and communicate multidimensional information may actually divert from the broader goal of trying to develop a solution to the issue. Furthermore, as noted by Bakkes et al. (1994), indices are specific to the process to which they pertain. There is no set of "universal indices." Thus, in an effort to concisely communicate data, there is the potential to distort or bias the meaning of the data, or to sacrifice precision for simplicity. We believe that the trade-off of simplicity need not be at the expense of accuracy, generality, or flexibility. In the next section we illustrate how one may effectively implement an index in data gathering and communication.

A Four-Stage Metamorphosis of Environnemental Indices

In the introduction we built the foundation for using an environmental index as a concise data communication tool. As we will illustrate throughout the chapter, environmental indices have evolved over time. We conceptualize the evolution of the environmental index in four stages:

1. *Simple Indices-* Early environmental indices were successfully utilized on specific data sets for focused environmental decision making and communication. While many of these indices have been widely implemented and accepted in the professional community, at times policy makers and analysts view the use of an index as a trade-off between simplicity and precision. In the first section of this chapter we provide an overview of several studies, projects, and government programs that have successfully utilized the basic environmental index for effective decision making and communication.
2. *Compound/Multicriteria Indices.* As simple indices became more common, the natural next step was to combine them. For example, the Conservation Reserve Program initiated in 1985 by the U.S. Department of Agriculture started as a simple

index to help the government retire cropland, primarily for the purposes of reducing soil erosion. At first the program targeted only soil erosion, citing other goals such as preventing sediment from entering water as secondary benefits. It was not until the implementation of the Environmental Benefits Index (EBI) that the government could aggregate and integrate disparate indices into a single, multicriteria measure, capable of addressing many environmental goals at one time. As a result of implementing the EBI, the CRP has been effective in not only reducing the incidence of soil erosion, but the CRP has also improved the environmental performance of these lands in general for water quality, wildlife, and governmental rental payment.

3. *The Impact Matrix.* As we will discuss throughout the chapter, one aspect that makes the environmental index particularly effective is that a large quantity of data is channeled into manageable snapshots of information. The goal is to communicate information comprehensively enough to support a well-informed decision, without inundating the target audience. Following this argument, the selection of information to be communicated highly depends upon the values and preferences of the decision maker, which may in turn, be highly influenced by contexts of the problem or environment, such as politics. In the next stage in the evolution of the environmental index, we present the impact matrix, which can represent information comprehensively without compromising the flexibility to later develop different types of indices for different users. With a foundation embedded in the finance literature, the impact matrix is able to accommodate many different data sets, as well as yield different visual formats for indices, depending upon what the user values and is considering during the decision making process.
4. *Disparate Stakeholder Management* –Although preferences between decision makers may actually be quite diverse, the previous applications that we have presented do not explicitly include stakeholder preferences. In the fourth stage in the evolution of the environmental index, through a technique called Disparate Stakeholder Management, we illustrate how environmental indices may be used for group negotiations and decision making by including indices about stakeholder preferences.

Each of these evolutionary stages is discussed in the following sections, along with contemporary examples of their uses. At the conclusion of this chapter, we summarize our findings, and suggest that the next frontier in the evolution of the environmental index may utilize the index as a tool to increase resource efficiency.

Simple Indices

The EPA and the NAAQS

Although indices such as the Dow Jones Industrial Average (DJIA) and the Standard and Poor 500 (S andP 500) have been utilized in economics for decades, the use of the environmental index is a more recent phenomenon, with very similar objectives. The most classic example of an environmental index arose from the assessment and monitoring procedures promulgated by the U.S. Environmental Protection Agency's (EPA) Clean Air Act and Clear Water Act in 1972 and 1976, respectively. With regards to the 1972 Clean Air Act, the EPA established the requirement for states to adopt a minimum National Ambient Air Quality Standard (NAAQS) for six pollutants: carbon monoxide, nitrogen oxides, sulfur oxides, hydrocarbons, photochemical oxidants, and particular matter.

In order to illustrate a region's compliance with the state implemented NAAQS, indices were initially developed to provide a synopsis of whether—and to what extent—a region remained below the maximum level for each of these six pollutants. In essence, most decision makers, such as elected government officials and local agencies, were more interested in the strategic results of NAAQS rather than the tactical details. More specifically, it was more important for these policy makers to identify WHETHER their region was in compliance with the NAAQS, rather than become inundated with the scientific details of WHY their region was not in compliance.

One example of an inaugural NAAQS regional index is presented in Figure 1. Figure 1 is divided into a top and a bottom section, which illustrate standard violations for 8 regional counties according to days per year. The bottom portion of Figure 1 reflects the binary EPA compliance system. Primary standards, set to protect public health, are shown with a clear box, and reflect a lower threshold. Secondary standards (or alert levels) are developed to protect against more systemic damage, such as damage to materials plants and animals, reduced visibility, and economic losses. Clearly, an individual without a technical background can easily recognize the frequency and severity in which several counties violated ambient air quality standards.

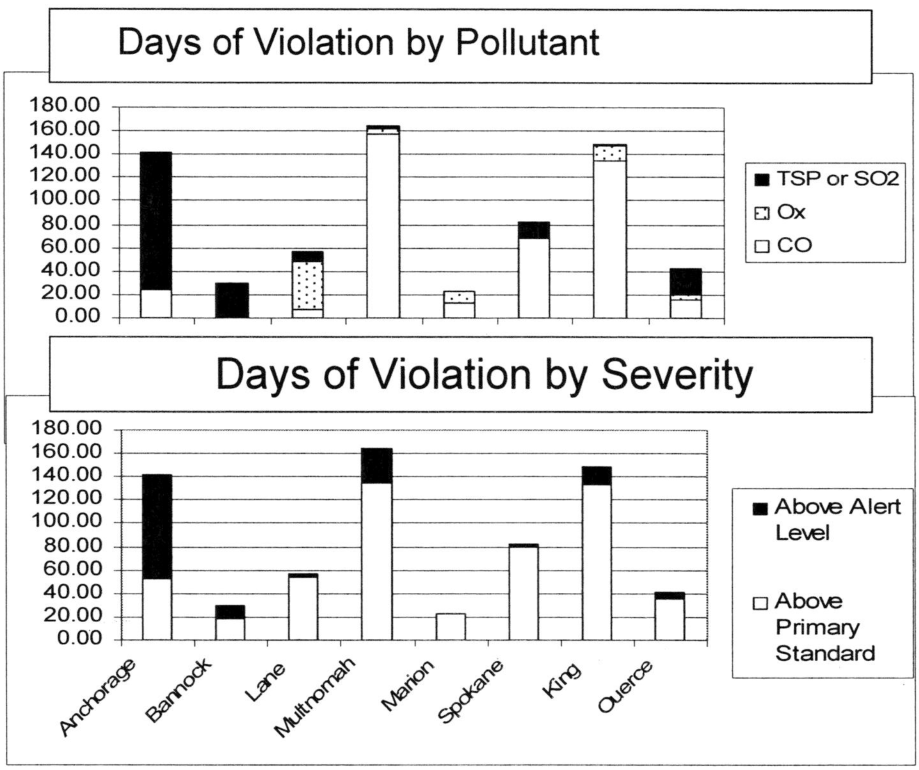

Figure 1-Regional NAAQS Compliance Indices (Adapted from Ott, 1978: p. 19)

Likewise, the upper segment of Figure 1 reflects the pollutants contributing to the air quality standards violations according to county. Once again, it is important to note that the decision maker does not necessarily need to know the precise level of the pollutant in parts per million; it is more important that the decision maker have an understanding of what pollutants may be triggering primary or alert level violations in order to develop an appropriate policy recommendation.

The NAAQS represents the most elementary example of an environmental index. It is an index utilized by one specific organization (the EPA) to communicate a very basic level of information—the National Ambient Air Quality Standards. However, this simplicity is not without its trade-offs. Based upon the above NAAQS index, a reader does not have insight about **how** to lower the pollutant content. The reader only sees **whether** the pollutants are under the minimum allowable standard. In addition, the reader cannot easily apply this index to other objectives—the index is only pertinent to National Ambient Air Quality data.

While simplistic and quite focused, the NAAQS is an example of an effective index from the perspective of a government policy maker because it supports a policy maker's goals of effective decision making and the communication of information. For example, rather than contemplate what compounds compose particulate matter in the NAAQS index, most policy makers would probably prefer to determine whether or not the level of particulate matter in their region is at a dangerous level.

To further illustrate this point, consider whether these policy makers actually need to have a thorough understanding of the seemingly small nuances of scientific data in order to make good public policy decisions and to communicate information. It is conceivable that, due to the potentially overwhelming amount of data, a comprehensive scientific report may actually obscure important facts to a policy maker who lacks training in this particular area of scientific specialization. As a result, being unable to distinguish the important information from the more trivial, the policy maker's decision making ability may be somewhat occluded and issues may not be properly prioritized; resultantly, internal and public communication may also be compromised. Thus, from the world of a policy maker, the NAAQS is an example of a well-designed index that fulfills the goals of decision making and communication rather well, in light of its simplicity.

Inarguably, the NAAQS is one of the earliest examples of an environmental index that has been effectively used and widely accepted by the mainstream. However, as stated above, its use is strictly confined to air quality standards, illustrating one of the differences between simple indices (such as the NAAQS) from more recently developed compound matrices. In addition, another important factor to consider is that values, preferences, and politics, while paramount, are often overlooked, and rarely quantified by the more primitive examples. Returning to the particulate matter example, a policy maker or constituent may require the following information in order to make a well-informed decision:

- What are the consequences of particulate matter readings in excess of NAAQS standards?
- Are these levels even relevant if they occur in a region that is uninhabited?
- In light of the consequences, is it a priority to reduce the level of particulate matter when there are other environmental issues within this region?

While a thorough scientific report may contain a healthy dose of data, Hyatt and Hoag (1997) assert that data-rich programs may actually fail when the information is not made policy relevant. A prime example of this problem, according to Hyatt and Hoag (1997) and Ribaudo et al. (2001) is the U.S. Environmental Monitoring and Assessment Program (EMAP), which was introduced by the EPA to revolutionize the process in which environmental decisions are made. EMAP was quite detailed in its scientific description of environmental assessments and ecological science. Nevertheless, after initial success, according to Hyatt and Hoag, the program ultimately failed because EMAP neglected to develop a process for transferring the scientific observations into policy recommendations or relevant action items. The authors believe that the source of EMAP's problems arose from the fact that even early on, EMAP did not appropriately factor how peoples' values affected the collection and interpretation of data.

Thus, the important factor to glean from this discussion is that, in contrast to an overwhelming scientific report, an index is able to succinctly summarize scientific points to communicate and provide policy objectives to aid in decision making.

OECD's Core Indicators for Environmental Performance Reviews

To provide a concrete example of the wide spread acceptance of the environmental index as an information communicator and a decision making tool, we next describe one of the most widely used set of indices for environmental policy making and decision making: The Organisation for Economic Co-operation and Development (OECD)'s Core Set of Indicators for Environmental Performance Reviews (OECD, 1993).

When the OECD Council met in 1989, the Council prioritized striking a balance between international environmental policy and economic development decisions. This desire to stabilize these often polarized goals was reiterated at the next following G-7 summits, and consequentially, an OECD Council Recommendation on Environmental Indicators and Information was developed in 1991, with the goal of collectively improving the world's performance in environmental management.

To paraphrase OECD's definition of an environmental index, an environmental index is an aggregate of indicators or weighted values, which describe the state of the environment directly resulting from the significant change of a parameter value. Specifically, the OECD utilizes a "Pressure-State-Response" framework to describe the pressures on the environment cause by human activities:

- "Pressure" is an indicator of human impact upon the environment.
- "State" is an indicator of the environmental conditions
- "Response" is the indicator of the societal responses

The OECD approach relies on a table like that shown in Figure 2 to present information about pressure, state and response for variables of concern by policy decision makers or stakeholders. In this example, the public is concerned about climate, eutrophication, acidification, toxic contamination, biological diversity and waste management, among other issues. These were taken directly from an OECD study that estimated indices for each of the variables.

	Pressure Indicators of environmental pressures	State Indicators of environmental conditions	Response Indicators of societal responses
Issues			
Climate Change			
Ozone Layer Depletion			
Eutrophication			
Acidification			
Toxic Contamination			
Urban Environmental Quality			
Biological Diversity			
Landscape			
Waste			
Water Resources			
Forest Resources			
Fish Resources			
Soil Degradation			
Other			

Figure 2 Illustration of an index structure with 14 environmental issues that are evaluated for policy and decision making: (Source: OECD, 1993: p. 12)

Compound/Multicriteria Indices

As discussed above, the basic environmental index is an effective tool for presenting data on a relatively focused topic. Early examples of environmental indices were consistent in this format, and utilized quite effectively, as supported by Lindsey, Willman and Rummel (1997), who noted that basic environmental indicators have been "used independently (and appropriately) when the environmental condition or phenomenon of interest is unidimensional or a narrow aspect of a problem is of interest." However, the authors point out that most "conditions…are multidimensional and two or more indicators are required to describe or represent them adequately."

Using two examples from the U.S. agricultural literature, we show that an environmental index can also serve as a template that may be used by various government officials and policy makers for decision making. By developing innovative methods to develop and represent the value weighting system, the index can be adapted to reflect changes in political preferences and decision making, as well as diverse data sets. We now turn to examples of environmental index templates drawn from U.S. agricultural programs that may be used to represent multidimensional decision making analysis.

The Environmental Benefits Index in the U.S. Conservation Reserve Program

According to Heimlich (1995) Congress initially established the Conservation Reserve Program (CRP) in 1985 as a "voluntary long-term cropland retirement program" to be administered by the U.S. Department of Agriculture (USDA). The CRP was first authorized by the Food Security Act of 1985, then reauthorized in 1990 and 1996. Since 1985, as many as 36 million acres (about 10 percent of U.S. cropland) at any one time has been included in the CRP.

In essence, CRP participants were compensated annual rents and half the cost of growing vegetative cover (normally grasses or trees) to retire land that the USDA determined to be highly erodible or environmentally sensitive cropland. Program participation was typically 10 years. The Food and Security Act of 1985 originally required that 40-45 million acres of land be enrolled in the program by the end of 1990.

Acceptance into the CRP was initially determined through a basic eligibility process. That is, any land meeting minimum program requirements was accepted into the program, as long as the program administrative costs were within a maximum acceptable rental rate (MARR). The first nine CRP sign up periods accepted land into the CRP according to this process.

According to Heimlich, paramount research by Richelderfer and Boggess (1988) suggested that the environmental performance of the CRP could be improved if the USDA were able to rank bids according to how well the bids successfully fulfilled the primary objectives of the CRP, including, but not limited to the prevention of soil erosion. Hence, the Environmental Benefits Index (EBI) was created to quantify how successfully a parcel of land was able to fulfill the program objectives, as well as minimize government costs.

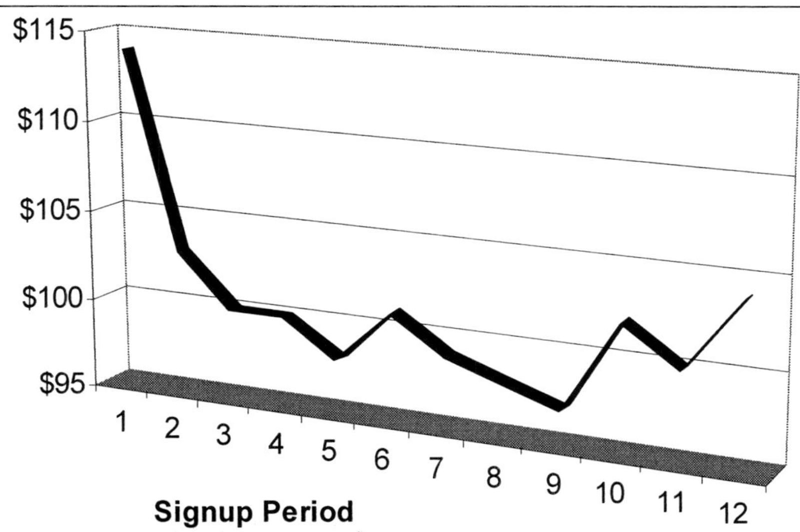

Figure 3: Environmental Benefits Index, by signup, for the Conservation Reserve Program (Source: Adapted from Heimlich, 1995: p. 102)

The EBI criteria were used to determine the three subsequent bids for acceptance into the CRP. Looking at Figure 3 from Heimich, it is not surprising that the initial CRP sign-ups illustrated high marginal value (as measured in environmental benefits per dollar), and subsequent enrollments showed declining marginal environmental benefits. It is very surprising, however, that following the official implementation of the EBI, subsequent parcels accepted into the CRP showed an increasing marginal environmental benefit per dollar.

Clearly, quantifying the CRP's environmental benefits, assisted through the use of an index, has broadened the program's benefits beyond the program's initial and primary focus of erosion reduction. Heimlich went on to create an index of environmental performance per dollar, and according to these results presented in Figure 4, the benefits from improvements in water quality and wildlife habitat were greater on a per-acre basis, than the benefits presented from reduced soil erosion under the CRP.

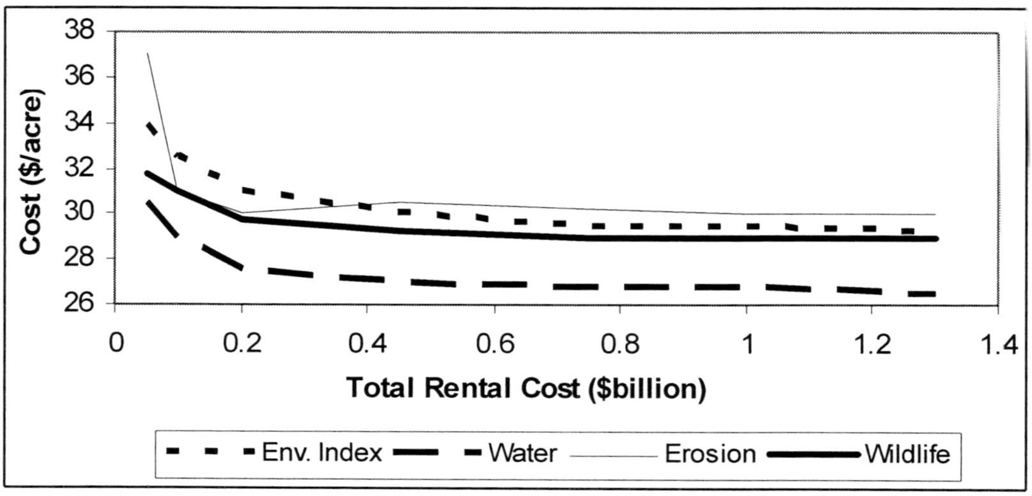

Figure 4: Environmental Performance per Dollar for the Conservation Reserve Program (Source: Adapted from Heimlich, 1995: p. 108)

The EBI as a Success Story

The CRP's positive generation of multiple environmental benefits created a successful feedback loop. As Ribaudo et al. point out, further refinements to the CRP enrollment process were made for the 1995 sign-up period to qualify land for the CRP based upon environmental benefits provided beyond the prevention of soil erosion. The authors note that the CRP was particularly found to improve wildlife habitat and that the EBI was further revised to evaluate the benefits to wildlife.

Significant in the evolution of the EBI was that for the first time the criteria were weighted to prioritize the most important factors in determining the outcome of the program. Water quality protection, creation of wildlife habitat, and soil erodibility earned the highest point values, while tree planting was also factored, but at a lower weight. The sum of these four criteria were divided by the cost factor (or the annual rent rate). These criteria were expanded in the 1996 and 1997 subsequent CRP enrollment process to include improvements in air quality due to reduced soil erosion and to give emphasis for land within a targeted environmental region.

More than 16 rounds of enrollments have been conducted for the CRP program since 1997, with additional weighting and criteria adjustments being made for additional environmental benefit contributions. The CRP's Environmental Benefits Index represents a success story in that the EBI initially quantified benefits per dollar, thus shifting enrollment away from marginal cropland to including more beneficial land in the program. Additionally, these indices revealed that additional environmental benefits were being generated from the retired cropland, thus expanding the policy making criteria and subsequently the policy and decision-making goals.

The CRP is considered a successful program, and the EBI has proven to be a living, breathing tool that has powered the engine behind the program's success. The fact that the EBI is capable of handling multiple data sets and has evolved to accommodate a weighting process places it as an example of the second stage in the evolution of the index.

As we will illustrate below, the advanced use of a weighting system provides the next stage in the progression of the environmental index.

THE IMPACT MATRIX

The impact matrix was introduced in a study by Hoag, Ascough, and Engler-Palma (2003). In this study, the authors present a matrix template capable of accommodating changes in user preferences, risk, and multiple data sets. As the authors illustrate, changes in preferences will adjust the index's weighting system, which may yield very different management and political decisions. The index allows developers to represent large amounts of data without making choices about which index will be displayed to a user. This has the distinct advantage of letting the ***users*** choose how they wish to trade off simplicity for complexity, instead of the index ***developer***. The matrix is not an index, but rather a structure from which to gather indices and to present them in different formats.

The authors borrow the concept of the payoff matrix from the financial risk literature to develop an "impact matrix", which they adapt to accommodate agroecosystem risk. The impact matrix essentially consists of a vector of plausible environmental indicators and outcomes for agricultural systems. Changes in these vectors yield the different weights, enabling the instrument to be adapted and used in a variety of contexts while remaining simple to understand. In addition, the impact matrix avoids several of the problems inherent to the weighting process, such as implicit weighting. The avoidance of implicit weighting also makes the impact matrix a very desirable decision making instrument.

Before discussing the impact matrix, it is first important to describe the theory behind the weighting process, then to discuss the payoff matrix. As Hoag et al. discuss in their research, the aggregation of variables into a single index requires an aggregator function, implying that variables should be subject to a weighting process. However, designing universal aggregator functions is difficult because the value judgments associated with assigning weights (called implicit weighting) are unavoidable. Furthermore, because values are highly subjective and variable, it becomes virtually impossible to develop aggregate weighting between cross-sectional groups (Hyatt and Hoag, 1997).

Hoag, et al. cite Heimlich's (1995) list of unavoidable choices that have to be made when creating indices: (1) apply explicit weights; (2) do not weight (implying distributional equity) (3) do not aggregate. However, they then move on to show that there are many shades of gray between these choices that can be borrowed from the risk literature.

While the application of weighting may prove to be quite challenging, the importance of selecting the most appropriate weighting process should not be underestimated. Using the Organization for Economic Cooperation and Development (OECD) Pressure State Response example, Hoag et al. contrast the impact of alternative weighting strategies. The effect of these different weighting processes and the subsequent indices (as well as their influence on the decision makers) are briefly discussed below.

The Pressure-State-Response framework is considered perhaps the most widely used international approach for developing indicators (OECD, 1993). Simply put, the Pressure-State-Response framework evaluates how external human pressures (such as air emissions or pollution discharges) impact the environment. Environmental pressure responses can be sub-divided into two variables: state variables, which are indicators of the quality of the environment, and response variables, which describe how society responds to these pressures.

Both state and response pressure variables have been evaluated by various groups and subsequently created into environmental indices. Figure 5 presents one example of an environmental pressure index that was developed by the Dutch government (Adriaanse, 1993) to measure state variables. In this case, target groups representing different human activities created environmental pressure on a set of eight themes. Pressure equivalents, the means by which a pressure variable is converted into a standard variable, transformed, and normalized, measures the contribution of each target group to each theme. Figure 5 shows an overview of pressure for the climate, acidification, eutrophication, and disturbance themes. Although transformed and standardized to target values for the year 2000, as seen in Figure 5, the presentation implies no weighting.

The OECD approach uses the first method described by Heimlich. That is, the developer makes assumptions about the user's weighting scheme. In contrast, another way to represent the same data is through the AMOEBA diagram (Heimlich's third method), as illustrated in Figure 6 (Brink et al, 1992). While the lack of weighting in Figure 5 shows a one-dimensional view of the data, alternatively, the AMOEBA diagram shows progress toward several goals simultaneously, without being confined to a single index. In this illustration, the authors have mapped the themes weighted by targets for simplicity. As is clearly shown in the AMOEBA diagram, the progress charted for years 1985 and 1990 (using 1980 values as a reference), shows that improvements were made to three themes (climate, acidification, eutrophication) while a fourth theme, disturbance, has become worse.

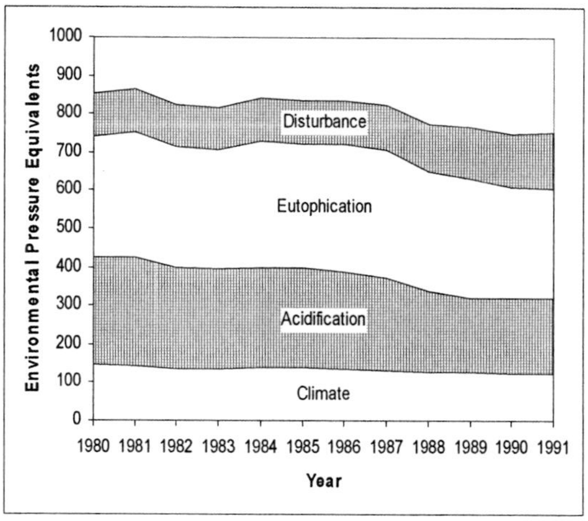

Figure 5: Environmental Index with No Weighting (Source: OECD, 1993)

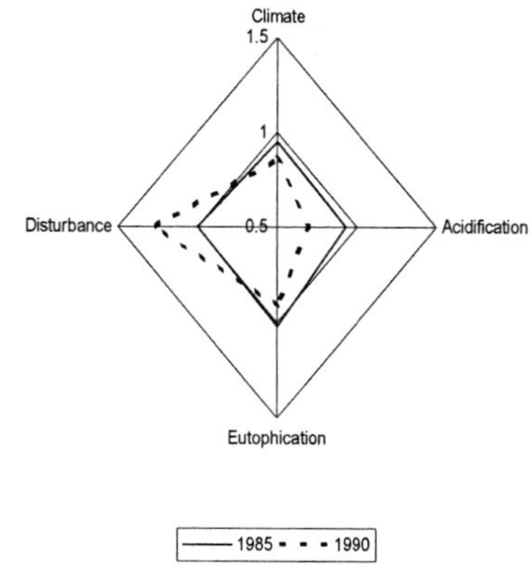

Figure 6: AMOEBA Diagram (Source: Brink, et al., 1992)

The advantage of presenting the weighting in the AMOEBA example is clear: a policy maker can easily review the data and determine that future work should focus upon improving the human disturbance on the environment without any pre-interpretation from whoever prepared the index. As illustrated with these two examples, the use of weighting can have a dramatic impact upon the information presented to decision makers; ultimately, the selection process and the right means for determining the appropriate strategy should be carefully considered. The AMOEBA is effective for simple problems and for decision makers who want to maintain some control.

While the AMOEBA approach accommodates multi-dimensional data, it lacks the capability to show multiple indices at one time. In addition, the AMOEBA diagram fails to consider risk and uncertainty, which may directly impact the respondent's preferences. By fulfilling these voids, the impact matrix, based upon the finance discipline's pay-off matrix, provides a distinct advantage over the previous two examples, while also overcoming problems of implicit weighting.

The Payoff Matrix

The payoff matrix is one of the most common tools used in decision theory. It is effective because the process quantifies the uncertainty involved with each outcome. Returning to the state-action variable, each decision involves:

- An analysis of possible actions
- The potential states of the environment
- The unique outcomes for each state-action combination

The unique outcomes are then arranged in a matrix called a payoff matrix, which enables a decision maker to utilize one of several potential decision rules. Applying the concept of the pay-off matrix to agroecosystem risk, Hoag et al. present the example of a farmer who is weighing a macroeconomic investment decision, in trying to determine whether to purchase new equipment or production facilities.

An example of the traditional pay-off matrix is presented in Figure 7. The first orientation is with the decision alternatives, which are listed on the vertical axis. The decision alternatives represent the magnitude of potential investment spending. The size of investment spending ranges from large [e.g., a new irrigation system (A1)] to moderate [e.g., a new combine (A2) or farm shop building (A3)] to no new investment (A4).

Next, the type of investment the farmer selects is dependent on future economic conditions, or economic states, which are represented on the horizontal axis. For example, will the price of grain the farmer sells be high [strong market (S1)], medium [fair market (S2)], or low [poor market (S3)]? Given the four decision alternatives and the three states of nature, which investment opportunity should the decision maker (farmer) choose?

Logically, in order to answer this question, the farmer needs to have complete information with respect to the profit associated with decision alternative/state of nature alternative. For example, how much profit would the farmer experience if he/she decides to purchase a new combine (A2) and economic conditions are good (S1)? This data is then organized into a payoff matrix, and hence, entries in a payoff matrix can be stated in terms of profits, costs, or any other measure of output that may be appropriate for the particular situation. Figure 7 below is the payoff matrix representation of the farm investment problem, where table entries signify profits for each state-action combination.

States/ Actions	S1	S2	S3	Maximax Criterion	Maximin Criterion	Minimax Regret Crit.	Hurwicz Crit.	LaPlace Crit.
A1	500,000 (0)	150,000 (50,000)	-150,000 (150,000)	500,000	-150,000	150,000	240,000	166,667
A2	300,000 (200,000)	200,000 (0)	-100,000 (100,000)	300,000	-100,000	200,000	140,000	133,334
A3	200,000 (300,000)	100,000 (100,000)	-50,000 (50,000)	200,000	-50,000	300,000	100,000	83,333
A4	0 (500,000)	0 (200,000)	0 (0)	0	0	500,000	0	0

Figure 7: The Payoff Matrix (Source: Hoag, et al., 2002)

One key advantage of the payoff matrix is that it is used to organize data but it does not yet impose a decision rule. And, since the concept has been around for quite some time, many innovative people have developed different types of decision rules that take advantage of the payoff matrix, which gives it remarkable powere and flexibility. In order to interpret the data and to analyze the farmer's decision, Hoag et al. apply several of these decision criteria, adapted from the finance literature for environmental settings, including: the maximax, maximin, minimax regret, Hurwicz, and LaPlace Criterion, respectively. The summary of these five criteria have been taken directly from the Hoag et al. publication.

Maximax Criterion

The maximax criterion is an optimistic approach. It suggests that the decision maker examine the maximum payoffs of alternatives and choose the alternative whose outcome is the best. This criterion appeals to the adventurous decision maker who is attracted by high payoffs. This approach may also appeal to a decision maker who likes to gamble and who is in the position and who is able to withstand any losses without substantial inconvenience. For the above example, the maximum among the maximum values is $500,000, so the decision is to purchase an irrigation system (A1). This value is found by first selecting the maximum (across the states) for each action, and then selecting the overall maximum from the "action maximums." That is, the best you can get for A1 is the best you can get for any system, given uncertainty.

Maximin Criterion

The maximin criterion is a pessimistic approach. It suggests that the decision maker examine only the minimum payoffs of alternatives and choose the alternative whose outcome is the least bad. This criterion appeals to the cautious decision maker who seeks to ensure that in the event of an unfavorable outcome, there is at least a known minimum payoff. This approach may be justified because the minimum pay-offs may have a higher probability of occurrence, or the lowest payoff may lead to an extremely unfavorable outcome. For the above example, the farmer would choose not to invest in new equipment or facilities (A4) because a payoff of $0 is the maximum among the minimum payoffs. That is, it protects the decision maker with the best income if the worst scenario happens.

Minimax Regret Criterion

The minimax regret criterion examines the regret, opportunity cost or loss resulting when a particular situation occurs and the payoff of the selected alternative is smaller than the payoff that could have been attained with that particular situation. The regret corresponding to a particular payoff is defined as Ras = Xs(max) – Xas where Xs(max) is the maximum payoff attainable under the state Ss (the subscripts a and s refer to the action and state, respectively). This definition of regret allows the decision maker to transform the payoff matrix into a regret matrix. The minimax regret criterion suggests that the decision maker look at the maximum regret of each strategy and select the one with the smallest value. This approach appeals to cautious decision makers who want to ensure that the selected alternative does well when compared to other alternatives regardless of what situation arises. It is particularly attractive to a decision maker who knows that several competitors face identical or similar circumstances and who is aware that the decision maker's performance will be evaluated in a relation to the competitors. This criterion is applied to the same decision situation and transforms the payoff matrix into a regret matrix (shown in parentheses in Figure 7). For the minimax regret criterion, the farmer would choose to purchase a new irrigation system (A1) because the payoff of $150,000 is the minimum among the maximum regrets.

Hurwicz Criterion

This criterion strikes a compromise between the maximax and maximin criteria. The decision maker is neither optimistic nor pessimistic (conservative). With the Hurwicz criterion, the decision payoffs are weighted by a coefficient of optimism α (realism), where $0<\alpha<1$. Conversely, $(1-\alpha)$ represents a measure of the decision maker's pessimism. The Hurwicz criterion requires that, for each decision alternative, the maximum payoff be multiplied by the coefficient of optimism, and the minim payoff be multiplied by the coefficient of pessimism. Assuming that the coefficient of optimism is .6, the weighted value for action A3 (i.e. purchase a farm shop building) in Figure 7 can be computed as: (US$ 200,000)(.6) + (US$-50,000)(.4) = US$ 100,000

In the above farm investment example, the maximum value for the Hurwicz criterion is US$ 240,000; therefore, the decision is to purchase an irrigation system (A1).

Laplace Insufficient Reason Criterion

The Laplace insufficient reason criterion, also called the equal likelihood criterion, postulates that if no information is available about the probabilities of the various outcomes or states, it is reasonable to assume that they are equally likely. Therefore, if there are n outcomes, the probability of each is 1/n. This approach also suggests that the decision maker calculate the expected payoff for each alternative and select the alternative with the largest value. The use of expected values (EV) distinguishes this approach from the criteria that use only extreme payoffs. This characteristic makes the approach similar to decision making under risk. In the above farm investment example there are three states; therefore, the (equal) probability of each state is .333. The equal likelihood weighted value for action A2 (i.e. purchase a combine) in Figure 7 can be computed as:

(US$ 300,000)(.333) + (US$ 200,000)(.333) + (US$ - 100,000) (.333) = US$ 133,334

The maximum value in Figure 7 for the Laplace Insufficient Reason criterion is US$ 167,667, so the decision is to purchase an irrigation system (A1).

Expected Value

It may also be possible to obtain probability estimates for each state of nature in decision making situations. If probabilities are known, then the expected value criterion can be used to identify the best decision alternative. The expected value (EV) is calculated in a fashion similar to the Laplace criterion; however, instead of equiprobable states, each decision outcome (payoff value) for each state of nature is multiplied by the probability of occurrence. Other criterions that incorporate probabilities with decision making include the "expected opportunity loss" (EOL) criterion and the "safety first" criterion.

To summarize, the farm investment example illustrates that use of several decision criteria often results in a mix of decisions. In the big picture, this pay-off matrix example effectively considers risk and uncertainty, which may yield very different outcomes, depending upon the criteria utilized by the decision maker. Ultimately, the appropriate criterion is dependent on the "risk" personality and philosophy of the decision maker.

From Payoff to Impact Matrix

In the next phase of the Hoag et al. article, the authors apply the premise of the payoff matrix to develop an environmental indicator. In order to do this, the authors construct an "impact matrix", based upon the OECD Pressure-State Response method discussed earlier in the chapter.

To illustrate the impact matrix, the authors utilize data from the U.S. Department of Agriculture to evaluate the potential pressure-response resulting from four possible soil management practices for a field near Tifton, Georgia, in the Southern Coastal Plain of the United States. As follows is a brief overview of the characteristics of the field for the current crop management practices (Considered to be the "Conventional" practice, and represented on the horizontal axis in Figure 8):

- Soils are Tifton sandy loam on 2% to 5% slopes.
- Drainage area is approximately 29 ha
- The main crop is continuous corn
- Fertilizer is applied annually with 115 kg/ha nitrogen and 30 kg/ha phosphorus

The alternative procedures, which will be represented on the horizontal axis, are:

- Alternative 1: Continuous corn and a small grain winter cover crop
- Fertilization method: 115 kg/ha nitrogen and 30 kg/ha of phosphorus are applied annually
- Alternative 2: A terrace system with grassed waterways, continuous corn and a small grain winter cover crop
- Fertilization method: 115 kg/ha Nitrogen and 30 kg/ha of Phosphorus are applied annually
- Alternative 3: Fair pasture
- Fertilization method: Single application each spring of 100kg/ha of Nitrogen

The most significant "pressures" that each of these alternative cropping systems may exert on the environment are:

- SY: Sediment yield (expressed in tonnes/ha)
- RO: Run-off (expressed in mm)
- NLG: Nitrogen leaching to groundwater (expressed in ppm)

The contribution of the four alternative field practices towards each of the three environmental pressures are quantified and presented in Figure 8 below:

Pressures/Methods	Conventional	Alternative 1	Alternative 2	Alternative 3
SY (tonnes/ha)	4.72	3.83	0.79	0.09
RO (mm)	80.07	58.22	30.17	14.12
NLG (ppm)	9.6	7.1	8.7	26.1

Figure 8: Matrix of environmental pressure variables

As we will show in the next paragraph, data contained within this impact matrix can be subjected to any of the five financial risk criterion discussed in the previous section (Maximin, minimax, minimax regret, Hurwicz, or Laplace criterion). Furthermore, the data contained within this impact matrix can be subjected to maximizing the expected value and the rule of insufficient reason, which is reflective of Figure 7. Furthermore, the AMOEBA diagram may also be used to represent the entire impact matrix itself, since all three impacts are listed in the matrix. Therefore, the impact matrix gives decision makers a tool that can be used depending upon different needs and objectives.

In order to evaluate the data and apply the various decision rules, as stated earlier in the paper, one must first convert the three pressure variables into a pressure equivalent indicator (EPeq). The EPeq pressure value is unitless, and can used as a comparison tool across variables, which would otherwise be expressed in different units. This enables us to develop a compound multicriteria index, allowing us to "add apples and oranges", so to speak.

The formula for calculating the $EPeq_{pressure\ variables}$=Impact (units)/target (units)

For the impact units, we can utilize the data collected in Figure 8 for each of the four alternative soil management strategies and the impact upon each of the three pressures. With respect to the target values, Hoag et al. utilized the value of 8 metric tonnes/ha as a proxy for the maximum sustainable value for erosion (Adriaanse, 1993), and USA health safety standards to measure maximum concentration level (MCL) for run off and nitrogen content.

Referring to the maximum sustainable erosion example, the EPeq values for the conventional system were:

$EPeq_{SYconventional}$= 4.72 (tonnes/ha)/8 (tonnes/ha) =.59

After subjecting each pressure to a neutralizing pressure equivalent indicator process, Hoag et al. generated an impact matrix of environmental indicators, which are presented in Figure 9. These unitless values may now be subjected to any of the previously discussed decision alternatives.

Pressures/Methods	Conventional	Alternative 1	Alternative 2	Alternative 3
SY	0.59	0.42	0.09	0.01
RO	5.64	4.12	2.14	1.0
NLG	0.96	0.71	0.87	2.61
Simple Average	2.40	1.75	1.03	1.21
Expected Value [1]	1.16	0.84	0.44	0.63
Maximax Value	0.59	0.42	0.09	0.01
Maximum Regret – SY	0.58	0.41	0.08	0
Maximum Regret – RO	4.64	3.12	1.14	0
Maximum Regret – NLG	0.25	0	0.16	1.90
Regret Index	5.47	3.53	1.38	1.90

*Assume that probabilities are the following: SY=.70, RO=.1, and NLG=.2

Figure 9: Impact Matrix for Selected Agricultural Management Practices

The recommended alternative depends upon which decision strategy is employed. The following observations can be made:

- For reducing adverse environmental impacts, Alternative 3 (fair pasture) is the best or second best management practice across all decision rules. For further explanation, please note:
    - For the simple average method, Alternative Three is the second best choice, at 1.21, compared to 1.75 (the best choice and 1.03 the second best choice).
    - In the Expected Value analysis, Alternative 3 is also the second best alternative, with .63, compared to Alternative 2 at .44.

  - With the maximax, the "best" of the "best scenarios" for contamination is to have the lowest level of contamination. Alternative 3 presents the lowest level of contamination at .01—the best maximax option.
  - The regret index, which summarizes the SY, RO, and NLG results, also opts for the lowest possible of the maximum regrets (i.e., minimax regret), so a low score is optimal. Alternative 3 presents the second lowest score.
- The minimax regret index is considered the preferred optimal solution, as it reflects the smallest value for the maximum possible regrets. These results have been summed in a "regret index".

DISPARATE STAKEHOLDER MANAGEMENT

The Hoag et al. study presents the impact matrix as a sophisticated tool to generate environmental indices that is reflective of the pay-off matrix from the finance literature. The impact matrix effectively communicates numerous data sets, thus enabling a user to employ a multitude of decision analysis strategies to determine the most effective approach for soil management. However, despite the sophistication of this tool, one should not lose sight of the second equally important objective for an environmental index, which is to communicate information in a concise manner. Indeed, Hoag et al. have illustrated that the impact matrix concisely communicates data to facilitate decision making, and it can easily be used by an economist or financial analyst trained in financial theory. However, it's flexibility is only exploited if the user knows what they are looking for. A user may be overwhelmed with too many choices.

The next step in the developmental process is to expand the applications of the IM to group negotiations and decision making. That is, stakeholders have to decide on how to use indices effectively when there are many indices to choose from. This will involve at least two questions, how to build biologically- or physically-based indices, and how to represent stakeholder preferences in those indices. Work by Lindsey et al. (1997), for example, showed that "systematic work to develop and use indices can elevate and focus the dialogue that occurs within an environmental planning process." This study was conducted with a fairly cohesive advisory committee to develop recommendations for a wellfield protection planning project in Indianapolis, Indiana. The procedure assisted City Planners in evaluating multiple alternatives. However, no mention was made of whether various stakeholder preferences were considered.

In contrast, a recent study by Caughlan and Hoag (2004) expands the implementation of environmental indices to a real decision problem in order to summarize the values and preferences exhibited by diversified stakeholders to promote group negotiations and inter-agency decisions. This study combined several decision variables by integrating stakeholder policy preferences with actual management alternatives to actually solve the problem so there was nothing left for the decision maker to answer.

This technique, known as the Disparate Stakeholder Management approach, was used to assist federal land managers in assessing diverse (and often polarized) stakeholder preferences in order to develop strategies to manage elk and bison herds in the Southern Greater Yellowstone area. Steps in the Disparate Stakeholder Management Process are as follows:

1) outline the big picture of the management problem, including resolution goals, through a hierarchal process. This typically involves the implementation of a questionnaire or survey;

2) after conducting the hierarchal process, the problem will then be reduced to smaller "decision points" that are being considered in the range of resolute actions; and

3) develop a policy preference map to summarize "decision points", and to quantify stakeholder preferences.

The policy preference map, another example of an index, will provide a visual representation of policy alternatives, which will concisely present decision makers with a greater understanding of the trade-offs between management decisions. Managers also obtain insight into how each stakeholder group might be impacted—or react to—these management decisions.

Study Specifics

In collaboration with the National Park Service and the U.S. Fish and Wildlife Service, Caughlan and Hoag collected the broad set of stakeholder preferences in the management of the elk and bison herd utilizing an Analytical Hierarchy Process (AHP).

The AHP approach organizes the management decision in a hierarchy framework that allows for the weighting of all factors influencing the decision allowing for a unitary, compromised solution. In the Caughlan and Hoag study, stakeholders were asked to complete a pairwise comparison survey to quantify their preferences for alternative herd management strategies and management goals.

From this raw data, a policy preference map was created to visually depict the continuum of stakeholder preferences and policy choices. A policy possibilities frontier (PPF), was developed to represent the technically efficient combinations of two types of land use management objectives with many alternative combinations of each objective possible along the curve that can be produced using available resources. As shown in Figure 10, the management options stakeholders weighted in the AHP preference survey and the actual management alternatives (potential policy options A-G) were placed within a PPF spectrum of management practices ranging from a "natural" land use approach to a more "managed" land use approach. Since the Analytical Hierarchy Process preference survey allowed stakeholders to choose any amount of each management option, some stakeholders fell between the alternatives. The visual depiction of stakeholder preferences enables the decision makers—in this case the National Park Service and the U.S. Fish and Wildlife Service—to quickly analyze the trade-off in values that each group places upon bison and elk management strategies.

This PPF provides a visual example of the newest generation of indices, which can facilitate inter-agency cooperation to attain a common goal. By following a somewhat methodical hierarchal procedure, social scientists may generate an inter-agency decision derived from the preferences and values of various organizations.

Relating the Caughlan and Hoag study to the overall focus of this chapter, Caughlan and Hoag have illustrated the next frontier of indices, which are implemented to achieve cooperation between disparate stakeholders in achieving environmental management objectives. This visually appealing policy preference map concisely illustrates the trade-offs between hands-off and managed land use practices, while representing a multitude of data points and a complex problem. And the entire system rests on a variety of indices.

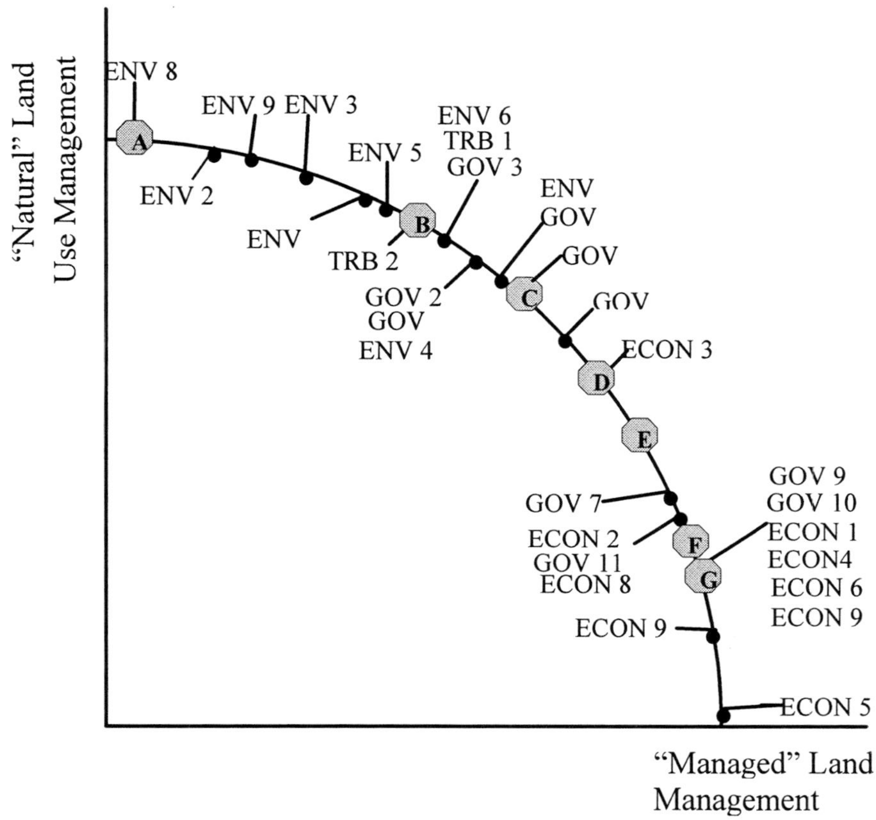

Organizational Code Key
GOV = Government Interests (includes federal, state, and local)
ECON = Economic Interests (includes agricultural, local business, hunting, and outfitting)
ENV = Environmental Interests (includesconservation groups and animal rights)
TRB = Tribal

Note: If more than one representative from an organization was interviewed, the combined preference scores are shown.

Figure 10: Elk Management Policy Possibilities Frontier (Source Caughlan and Hoag, 2004)

The Next Frontier of the Environmental Index Increasing Efficiency

Throughout this chapter we have built the case for the environmental index as a useful tool for communication and decision making. Since its inception in the 1970's, we have illustrated that the environmental index has evolved from a one-dimensional tool that simply and precisely presents data, into a flexible instrument, capable of accommodating multiple data sets, adjusted weights, and being used to represent multiple stakeholder interests.

Another aspect that we believe the index can be further developed to achieve is to increase resource efficiency. As illustrated in the Caughlan and Hoag study, the policy preference frontier reflects the trade-offs between "hands-off" land use management and "planned" land use management. Each stakeholder's interest is represented on the frontier as an efficient balance between the two land management options. No stakeholder interest is "infeasible"; that is, all stakeholder preferences represent a feasible balance between the two land management positions, and presumably, policy conditions have been developed to meet conditions to which each stakeholder will agree. Similarly, all interests lie exactly on the curve, meaning that the most efficient solution is being obtained.

The idea of utilizing an index to create efficiency was suggested by Ganoulis (2001), who uses a simple mathematical programming model. His research recommends an index called "Multicriterion Decision Analysis (MCDA)" problem to determine the most appropriate water pricing strategy in the Euro-Mediterranean agriculture system. With this method, the MDCA factors economic, technical, environmental and social aspects of water resources management are calculated with respect to a constraint. In this procedure, Ganoulis defines objectives, attributes, and constraints.

The *objective function* represents the desired change that a decision maker (in this case a policy maker) recommends for a system. Objectives may be maximized, minimized, or maintained. For example, in the case of water use management, the objective may be to maintain a specific level of downstream water. An *attribute* is the method for evaluating how an objective is obtained or measured; in this case, Ganoulis lists flood damage, sediment yield, and nitrate concentration as examples of attributes. A multi-attribute decision model, such as the one that Ganoulis employs, selects the best of a number of alternative alternatives, which are all described in terms of attributes. The *constraint* is the adherence procedure by which a policy maker must adhere. Laws, budgets, and capacity are all examples of constraints.

Ganoulis develops his model by seeking a non-dominated solution that improves all objective functions, rather than simply attaining an optimum solution. Using a frontier approach similar to Caughlan and Hoag, as shown in Figure 11, two alternatives (Y_1 and Y_2) are being compared. Points drawn along the curve are defined by a maximum value of objective function Y_1 given Y_2. This curve thus represents optimal solutions in the sense that there is no increase in a single objective function without a decrease in the value of the other objective function, and the solution is efficient. Any option located within the interior may be considered feasible but inefficient, while any option to the right of the curve is infeasible.

Ganoulis also illustrates the indifference curves into his model, which represent societal and policy preferences. Following the basic economic axioms, the indifference curves are monotonic, and the intersection with the frontier represents the most efficient solution for the problem. Ganoulis' then creates a composite ecological index and a composite economic index to find the most efficient solution, subject to different objective functions, alternatives and constraints.

Ganoulis's model could also be applied to the previously discussed Conservation Reserve Program, which yielded different objective functions than its original objective on soil conservation. Likewise, a similar approach with the Environmental Monitoring and Assessment Program (EMAP) may have generated more focused policy recommendations and a better fate.

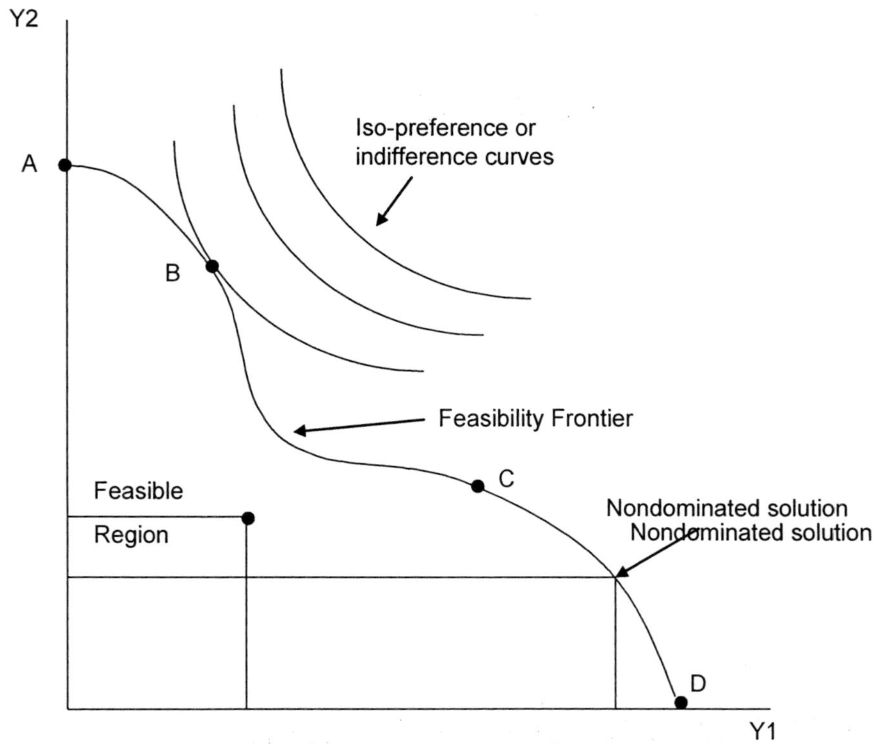

Figure 11: Non-dominated solutions for a two-objective problem (Source: Ganoulus, 2001)

CONCLUSION

In conclusion, we have demonstrated that the environmental index is a concise policy tool for data communication and decision making. In its simplest form the environmental index can encapsulate focused data, but the more sophisticated models are also capable of considering multicriteria data sets, risk, probability, and weights. Furthermore, the environmental index has been used to accommodate user preferences, and to facilitate decision making among disparate stakeholders. Additional applications of the environmental index may be used to increase efficiency, and the environmental index may serve as a tool for communicating solutions generated by mathematical programming models.

As we have shown throughout the chapter, policy makers may desire to use the environmental index because its simplicity assists with timely, appropriate decision making. However, as we have also illustrated, the environmental index is also a flexible tool that provides precision and efficiency in addition to—not rather than—simplicity.

REFERENCES

Adriaanse, A. (1993). Environmental policy performance indicators: A study on the development of indicators for environmental policy in the Netherlands, *SDU, Uitgevereji*, The Hague.

Brink, B., Hosper, S., Colin, F. (1991). A quantitative method for description and assessment of ecosystems: the AMOEBA approach. *Marine Pollution Bulletin*, 23, 265-270.

Caughlan, L., Hoag, D. (2004). Feeding Elk in Greater Yellowstone: A Case Study in Ex-ante *Group Decision Support*. (fourthcoming)

Ganoulis, J. (2001). Ranking alternative strategies of agricultural water use in the Euro-Mediterranean area. Round Table #7, *Euro-Mediterranean Forum of Economic Institutes*, Marseille, March 29 and 30, 2001.

Heimlich, R. (1995). Environmental indicators for U.S. Agriculture, In: S. Batie (Ed.), Developing indicators for environmental sustainability: The nuts and bolts, SR-89, Michigan State University Press, *East Lansing, MI*, pp. 99-116, 1995.

Hoag, D.L., Ascough, J.C., Engler-Palma, A. (2002). The impact matrix approach and decision rules to enhance index dimensionality, flexibility and representation. *Ecological Indicators*, 2, 161-168.

Hyatt, D.E., Hoag, D.L. (1997). How are we managing? Environmental condition is value-based: A case study of the environmental monitoring and assessment program. *Ecosystem Health*, 3, 120-122.

Lindsey, G., Wittman, J., Rummel, M. (1997). Using indices in environmental planning: Evaluating policies for wellfield protection. *Journal of Environmental Planning and Management*, 40(6), 685-703.

OECD (1993). OECD core set of indicators for environmental performance reviews, *Environmental Monographs.*

Ott, W.A. (1978). Environmental Indices: Theory and Practices. *Ann Arbor: Science Publishers Inc.*

Ribaudo, Marc, Dana Hoag, Mark Smith and Ralph Heimlich. Environmental Indices and the Politics of The Conservation Reserve Program. *Ecological Indicators*, Vol. 1 (2001): p. 11-20.

Richelderfer, K., Boggess, W.G. (1988). Government decision making and program performance: The case of the Conservation Reserve Program. *American Journal of Agricultural Economics,* 70(1), 1-11.

USDA-SCS, *User's Guide for CREAMS Model*, U.S. Department of Agriculture, Soil Conservation Service, Engineering Division, *Technical Release* 72, 632 pp., Washington, D.C., 1984.

U.S. General Accounting Office (1993). Conservation Reserve Program: Cost-Effectiveness is Uncertain. *GAO/RCED*-93-132, Washington, D.C.

Wildlife Management Institute (1994). The Conservation Reserve Program: *A Wildlife Conservation Legacy*. Washington, D.C.

In: New Trends in Ecology Research
Editor: A. R. Burk, pp. 183-227
ISBN: 1-59454-379-8

Chapter 8

ENVIRONMENTAL MANAGEMENT PROBLEMS, FUTURE GENERATIONS AND SOCIAL DECISIONS[1]

Joan Pasqual and Emilio Padilla

Department of Applied Economics, Universitat Autònoma de Barcelona,
Edifici B, 08193, Bellaterra, SPAIN

ABSTRACT

The decisions of many individuals and social groups, taking according to well-defined objectives, are causing serious social and environmental problems, in spite of following the dictates of economic rationality. There are many examples of serious problems for which there are not yet appropriate solutions, such as management of scarce natural resources including aquifer water or the distribution of space among incompatible uses. In order to solve these problems, the paper first characterizes the resources and goods involved from an economic perspective. Then, for each case, the paper notes that there is a serious divergence between individual and collective interests and, where possible, it designs the procedure for solving the conflict of interests. With this procedure, the real opportunities for the application of economic theory are shown, and especially the theory on collective goods and externalities. The limitations of conventional economic analysis are shown and the opportunity to correct the shortfalls is examined. Many environmental problems, such as climate change, have an impact on different generations that do not participate in present decisions. The paper shows that for these cases, the solutions suggested by economic theory are not valid. Furthermore, conventional methods of economic valuation (which usually help decision-makers) are unable to account for the existence of different generations and tend to obviate long-term impacts. The paper analyzes how economic valuation methods could account for the costs and benefits enjoyed by present and future generations. The paper studies an appropriate consideration of preferences for future consumption and the incorporation of sustainability as a requirement in social decisions, which implies not only more efficiency but also a fairer distribution between generations than the one implied by conventional economic analysis.

[1] Financial support from projects BEC2003-08131, Ministerio de Ciencia y Tecnología, and 2001SGR-160, Direcció General de Recerca, is gratefully acknowledged.

Introduction

In order to consider the problems arising in the management of natural resources and the environment it is first necessary, to describe the various types of goods involved in economic terms. The classification presented in section 2 starts from the classical divergence between private and collective goods and it is expanded with new concepts that give rise to new types of goods such as, for example, the ones that we call *subprivate* goods, *hypercollective* goods, and *intergenerational* goods. The conceptual framework is therefore enhanced and the conditions for a rigorous analysis are improved. It is also necessary to refine the conceptual basis provided by economic theory on *externalities* in order to analyze not only the classical divergences between individual and collective interest, but also the opportunities for action in a non-ideal context. These questions are considered in section 3 and the conflict that exists between present and future generations is analyzed in detail, from the perspectives of both equity and efficiency.

Conventional methods of economic evaluation of projects and policies are usually used to help decision-makers. The application of these methods to intergenerational problems, such as climate change, has been widely criticized because it does not take the interests of future generations into consideration adequately. Section 4 analyzes the limitations of conventional analysis of intergenerational problems and proposes an alternative solution which incorporates the existence of different generations and the requirement of sustainability in long-term projects. Section 5 studies an intergenerational weighting showing social preferences, and proposes an evaluation methodology which aggregates the costs and benefits of projects through this weighting. Section 6 analyzes the individual and social optimality of intergenerational investments from an individual and social point of view using an overlapping generations model with intergenerational altruism. Finally, section 7 summarizes the main conclusions.

Collective Goods and Other Peculiarities

Definitions and Classifications

The notion of *collective good* emerged as a response to a legal action against Thomas Jefferson for using the idea of a mechanism that was patented. The first example of collective good was therefore an idea. Jefferson argued that one should not pay for the use of an idea because "*its peculiar characteristic is that nobody possesses less of it due to the fact that another posses it completely*", i.e. because there is no any opportunity cost for anybody. The economic definition of *collective good* comes from Samuelson (1954). It is opposite to that of *private good* because of its consumption characteristics and it has no relationship with the type of ownership of the good. For a good to be *collective*, it does not matter whether the ownership is public, private or mixed.

In a *private good*, such as apples or tissues, there is rivalry in its consumption because if a unit is consumed, the total availability for consumption diminishes in this unit. However, in the case of a good with collective good characteristics, such as the protection provided by the ozone layer, all consumers consume the same amount — all the available quantity. It is obvious that the environment and natural resources clearly have collective good characteristics.

In a *pure collective good*, consumption is unavoidable, the cost of access is zero, and there is no possibility of excluding any consumer. This creates serious problems for financing the provision of this type of goods. In fact, given that nobody can be excluded from its consumption, even if they have not contributed at all to financing the provision of the good, there are no incentives for contributing to financing the provision of the good, which is known as *free rider* behavior. The same problem arises when the aim is to discover a consumer's valuation of the good. The individual will tend to undervalue (overvalue) the good if he/she thinks that his/her fiscal contribution will be greater (smaller) as greater (smaller) is the valuation stated. A prior requirement for efficient provision is knowledge of the good's economic value, the demand curve. As a consequence, in order to achieve an optimal provision, a demand revelation mechanism should be used, i.e., a procedure that induces consumers to tell the truth in their own interest.

Local Collective Goods

Goods that have collective good characteristics for only a subset of consumers and not for all of them are called *local collective goods*. These goods behave as pure collective goods *within* their range and as pure private goods *between* different ranges, such as the services that provide the clock of a church tower.

Hypercollective Goods

If one good is a collective good, the incorporation of an additional consumer does not affect consumption by other consumers. This study proposes calling goods whose available consumption increases when the number of consumers increases *hypercollective goods*. The utility obtained from the consumption of discotheques depends on the quantity of consumers. The same happens in certain types of associations, unions, political parties and groups of persons with similar preferences on specific issues, which provide the *relational goods* defined in Uhlaner (1989) –because the utility of each member increases with the total number of them.

Congestion

All goods with collective good characteristics can be affected by *congestion*. This means that when the quantity of consumers increases above a certain critical threshold, it is necessary to increase production in order to keep the characteristics of the good unchanged. For example, when a bus is full, it is necessary to use another bus to satisfy an additional user of this local collective good. Furthermore, in order to maintain the quality of a hypercollective good such as the Internet, it is necessary to enlarge its capacity when a congestion threshold is achieved.

Subprivate Goods

In a private good, a unit consumed reduces the total quantity available for posterior consumption by exactly this unit. We propose calling those goods whose available consumption decreases by more than one unit when a unit is consumed *subprivate goods*. The typical example is the use of the urban underground system by networks of collective services – see Pasqual and Riera (2004) - given that the space actually occupied is much lower than the total space disabled for other uses. Furthermore, depending on the fishing technique employed, for each ton of fish captured useful for human consumption there is a loss of much more tons of marine life. In order to consume one actual unit, it is necessary to produce K units (K > 1) of the subprivate good. Consumption efficiency (K) can be measured by observing how the total quantity available for consumption falls as a consequence of the actual consumption of a unit of good:

$$K = - \Delta \text{ available quantity} / \Delta \text{ consumed units}$$

The goods can be characterized, according to the degree of their private/collective characteristic, using the K factor of efficiency in consumption:

$$\text{Reduction in available quantity} = K \cdot (\text{quantity consumed})$$

The values of K for each type of the characterized goods are:

Type of good	K value
subprivate	$K > 1$
pure private	$K = 1$
local collective	$0 < K < 1$
pure collective	$K = 0$
local hypercollective	$K < 0$
hypercollective	$K < 0$

Durable Goods

A good is durable if it is not exhausted in only one act of consumption, when continuous use during a certain time period is possible and the various consumption acts take place before there is an appreciable deterioration of its basic characteristics. The characteristic of durable good is compatible with that of private goods – such as ties and lighters - and with that of collective goods – such as the force of gravity, the English language and the theorem of the inverse function. Unlike durable goods, non-durable goods are those goods whose consumption leads to the immediate destruction of the good, such as peanuts. A distinction should be made between those that are recyclable and those that not. A can of beer can be recycled, i.e., the material can be reclaimed at a certain cost. However, the use of oil as fuel completely destroys it for economic purposes, with no possibility of recycling it.

Intergenerational Goods

An especially interesting case of a durable good is all those goods with a duration of two or more generations, and those of unlimited duration in particular. For these types of goods, we suggest the name of *intergenerational goods*. The set of goods that constitute a culture and renewable goods are typical cases, not only of collective goods, but also of intergenerational goods, given that their benefits can be extended to an unlimited quantity of generations.

Intergenerational Collective Goods

Extending the definition of collective good to the intergenerational level, we obtain *intergenerational collective goods,* which are characterized by the feature that consumption by one generation does not at all limit consumption by the following generations. As a consequence, these goods are by definition sustainable. The characteristics of *intergenerational collective goods* are compatible with those of collective goods, local collective goods, private goods and any others within one generation. For instance, a jewel is a pure private good, a microclimate is a local collective good, the ozone layer is a pure collective good and Internet is a hypercollective good, but all of them are *intergenerational collective goods.*

Renewable Goods

A good is renewable, in the strict sense, when it is possible to have at one's disposal a certain minimum quantity of the resource for an unlimited time, with no need for human intervention because there is an automatic regulation of incomes and outcomes. The water of an aquifer, fishing resources, forests, and the stork population are examples of renewable goods. Logically, the characteristic of renewability does not entail the characteristic of indestructibility - it is easy to overexploit a forest or an aquifer for a short space of time or to annihilate the storks, destroying these flows of wealth forever. If biological goods are consumed at under their reproduction rate, the total available good for the future does not diminish because the mass recovers quick and automatically. However, the overexploitation of this type of resources entails the reduction of the quantity available for the future, and in extreme cases, the destruction of the resource. The non-biological ones, also named *environmental goods*, such as air, ozone layer, or oceans, do not reproduce and have a limited capacity for autoregeneration. If use does not surpass the limit of autoregeneration they could be used indefinitely as if they were *free goods* (goods whose opportunity cost is zero), while the result of overexploiting them is the continuous loss of quality.

Transformable Goods

Here we propose calling those goods where it is possible to change or modify their subprivate, private, collective, hypercollective, intergenerational, durable or non-durable characteristic *transformable goods*. A natural space is a local collective good but it can be used as if it were a pure private good if access is restricted to one person only. An art collection can be consumed as a collective good if it can be observed by any consumer, as a private good if it can only be enjoyed by its owner, and as a subprivate if it rests forgotten in the basement of a museum. The degree of transformation is reflected in the modification of the K factor of consumption efficiency. In addition, the duration of the goods is also transformable in general. It is as possible to ensure the survival of species that would disappear without human intervention as to irreversibly destroy a renewable good. There are multiple intermediate possibilities because the duration of a good can be modified in one or more periods. The duration of manufactured goods is a control variable for the maximization of private benefits by the producer, the chosen duration is systematically lower than the optimal[2] (planned obsolescence) and entails important external costs due to the increase in the use of materials and the generation of residuals. Increasing the duration of a manufactured product constitutes then a good measure of environmental policy, because the greater the

---

[2] In the absence of regulation, the duration of the products is lower than the optimal level and the quantity provided is inefficient due to excess. This is due to the fact the external cost produced by the residual outputs is not taken into account. The optimal duration and, at the same time, the optimal quantity produced and consumed, can be achieved through a tax a la Pigou. See Ocaña and Pasqual (1994).

useful life of one good, the lower the use of materials and the quantity of residuals produced. A mild transformation of the characteristics of a collective good, in order to make it less pure, has its advantages. An example is the restriction of the access to a natural park, which may be a necessary measure for the preservation of this space. In other cases, it is useful to remove the characteristic of non-exclusion in order to make private provision of a collective good feasible, and thus avoid the possible financial constraints of the public sector.

A Practical Case of Optimal Provision of Collective Goods: An Insurance of Water Availability

There are many ways to analyze "water" from an economic perspective. Initially, water can be seen as a raw material good, a renewable natural resource that is contained in multiple deposit sites. Economically, human beings derive welfare from the use of water, either as an input in the production of goods and services, or through its final consumption. Usually, water with economic value has to be stored, transported, treated, or evacuated with some additional treatment. Often, all of the above are required at the same time. One way to increase the amount of economically usable water is to reuse it.

Since water is a useful resource (it provides welfare), it will be more valuable as it becomes scarcer. Prices should therefore reflect the relative scarcity of the resource as an economic good. Again, there is the possibility of increasing the amount of economic water in a given system, by enlarging the storage capacity or accessing new deposits, by extending the transportation infrastructure, by de-salting sea water, or by extending the life cycle of the economic water, for instance. The life cycle can be extended by re-using water, after treatment, for higher economic value activities (e.g. agricultural use) than the water would otherwise have been used for (e.g. returning it to a river ecosystem). "Producing" more water implies an extra cost, but also an extra benefit. A cost-benefit analysis could evaluate whether the extra benefit is worth the extra cost or not. To do so, the value of water – its shadow price - has to be approximated. This chapter deals mostly with the problem of how to estimate this water value and how to implement a cost-benefit analysis of water reuse.

A Naïf Model

Given people's preferences and resources, the quantity demanded of good x, e.g. water, depends on its price p. The direct demand function is therefore $x = X(p)$. The price is used as a value signal and ought to coincide with the marginal cost. If p were not appropriate, the value signal would be biased and the amount of the good x provided would not be optimal. Instead of using the direct one, the indirect demand function could also be used, $p = P(x)$. The function $P(x)$ expresses the maximum price p the consumer would be willing to pay for an additional unit of water, if this is the good under consideration. In other words, $P(x)$ measures the marginal valuation of water by the consumer.

As an example, consider the demand schedule in table 1. If consumption were of 2 units of water, the (gross) social benefit would be 25 (19 monetary units for the first unit of water, and 6 for the second), while with 3 or more units the (gross) benefit would always be 28 monetary units.

Table 1. Actual marginal value per unit of water consumed

Physical units	1st	2nd	3rd	4th
Marginal valuation	19	6	3	0

Let us now suppose that the cost of providing water is that shown in table 2. With the data shown in both tables, the optimal consumption – the one that maximizes the (net) social benefit - is 3 units. The net benefit, or *surplus*, associated with the 3 units of water is $E^* = (19 + 6 + 3) - (1 + 1 + 2) = 24$ monetary units.

Table 2. Marginal cost per unit of water consumed

Physical units	1st	2nd	3rd	4th		Tth
Marginal cost	1	1	2	2	2	2

The result is the same as the one which would arise from a perfectly competitive market, in the absence of externalities, collective goods and increasing economies of scale. But the availability of water for consumption has collective good characteristics, and it is therefore susceptible to the free-rider problem. To illustrate this point, imagine that consumers understate their interest in the good. Instead of revealing the true demand schedule (table 1), they send signals according to table 3. With the cost schedule from table 2, the new "optimal" provision of water would be 2 units, since it is the amount that maximizes the social surplus $E^* = (19 + 6) - (1 + 1) = 23$ monetary units. There would therefore be a loss of welfare associated with the understated signals.

Table 3. Understated marginal value per unit of water consumed

Physical units	1st	2nd	3rd	4th
Biased statement (-)	12	5	1	0

If instead of understating their interest, consumers overstate it, the solution is again not optimal. To illustrate this, consider the new schedule in table 4. Now the "optimal" consumption would be 5 units of water, since this is the amount that produces the maximum surplus $E^* = (19 + 6 + 3) - (1 + 1 + 2 + 2 + 2) = 20$ monetary units. In conclusion, to achieve an optimal assignation of water, the free-rider problem has to be avoided.

Table 4. Overstated marginal value per unit of water consumed

Physical units	1st	2nd	3rd	4th	5th	6th
Biased statement (+)	20	17	8	4	3	0

There are still more problems to be aware of. The supply side might not be competitive. Let us consider for example a market with a single firm providing water to the locality, and say the firm is public and interested in enlarging its budget. Assume that the only restriction of the firm is to be profitable. In this framework, the provision will be the largest possible, while the surplus is strictly positive. Taking the true demand revelation of table 1as an example, the amount of water provided would be 14 units, and the surplus would be lower than optimal: $E^* = (19 + 6 + 3) - (1 + 1 + (12 * 2)) = 2$ monetary units.

In short, for the water projects to be efficient, the strategic behavior of both consumers and providers (among other problems) has to be avoided. To ensure this, all agents should have incentives to reveal their true preferences and behave efficiently. This is the subject of the next section.

The Proposed Model

Let water be demanded for both production and final consumption purposes. Suppose that the total amount of disposable water is usually sufficient. However, with a given probability, there may not be enough water to always satisfy both types. In deficit situations producers and consumers that demand water bear a cost associated with the shortage. In order to reduce the cost, an investment plan to increase water availability through a desalination plan is proposed. The unitary investment costs are known, but in order to determine the scale of the project, its (gross) benefits are to be estimated.

Consider that access to desalinated water has the characteristics of a collective good, and therefore does not offer individual incentives to contribute to its financing. All the demand agents are interested in the project being undertaken, but the free-rider problem arises. If individual agents are asked for their willingness to pay for the project, they will tend to understate (overstate) their value if they believe the answer would influence the amount of taxes they will have to pay. But the estimated demand function is drawn from the individual answers of the agents, and from it the (gross) social benefit is computed and the optimal amount of water determined. A procedure that ensures that it is in the best interest of each individual agent to honestly reveal his/her maximum willingness to pay would overcome the problem. There are several such procedures. The one proposed here is a variant of the Thompson (1966) mechanism.

The Demand Side

When the provision of one collective good has to be chosen from two or more alternatives, each agent may prefer one over the others, and their choices do not necessarily coincide. When the socially chosen alternative is not the one that a given agent prefers, he/she therefore bears an opportunity cost. However, as shown below, an insurance mechanism gives incentives to individuals to reveal their "true" opportunity cost. In this way, the demand revelation problem can be overcome, because the good that provides the largest social benefit is also the one that provides the minimum social opportunity cost.

To exemplify the problem, suppose one wants to find out the individual's maximum willingness to pay for a collective good, such as desalinated water in case of drought, for each potential consumer (or a sample of them) - in other words, to find out the individual and social losses of a drought season. Let us say that the government makes an announcement saying that it is prepared to offer insurance against drought (or water scarcity in general) to those who want to purchase it. The government informs that the probability of not having scarcity (event P) is p, while the drought (event Q) has a probability (1-p) of happening. For the mechanism to work properly, agents have to believe that the announced probability is the actual one. The unitary price to be paid for contracting insurance coincides with the drought probability (1-p), and at that price agents can insure for the amount of money they want. As shown below, this mechanism gives agents incentives to subscribe to the insurance for a quantity equal to the expected loss.

Assume the actual value (a, b, …) of each agent (A, B, …) is that expressed in table 5, where the aggregate willingness to pay is $V'(\cdot) = a + b + ... + n$. If agent A does not subscribe to the insurance (no S), his/her mathematical expected benefit (E_A) is the probability of having water (p) multiplied by his/her benefit (a) from using the water:

$$E_A \text{ (no S)} = pa \tag{1}$$

Table 5. Actual valuation of (desalinated) water during a drought

Agents	Valuations
A	a
B	b
…	…
N	N

If agent A contracts the insurance against drought, he/she has to decide the optimal amount in monetary units to receive in compensation (a + a'), where a' is the bias –or lie - introduced by agent A when revealing his/her valuation. When a' > 0, agent A overstates his/her interest; when a' < 0 he/she understates it; with a' = -a, the amount of the insurance is zero (no S); and finally, with a' = 0 agent A reports his/her actual (unbiased) valuation. The insurance premium will be

$$H_A = (1 - p)(a + a') \tag{2}$$

where (1-p)a' is the overpayment (or underpayment) of premiums due to the lie a' by agent A concerning events P (no drought) and Q (drought). The compensation payment for loss if Q occurs is the insured value (a + a'). The net benefit agent A obtains depends on whether P or Q occurs, which happens at probability p and (1 – p) respectively. The mathematical expected benefit for agent A, ($E_A(S)$), if he/she takes out insurance is therefore

$$E_A(S) = pa + (1 - p)(a + a') - H_A \tag{3}$$

And substituting H_A from (2),

$$E_A(S) = pa \tag{4}$$

An initial conclusion from (4) is that the expected net benefit to an individual agent from the insurance is regardless of the lie a'. Furthermore, (4) coincides with the expected benefit without insurance, (1). However, there is an important difference. The *risk* in subscribing to the insurance at the honest valuation is different to subscribing to it at a biased valuation, or in not subscribing to it at all.

Without insurance, the agents will obtain either the value a or zero, depending on whether P or Q occurs, which depends on the risk of drought, which might not be negligible.

Furthermore, if agent A gets insured against Q, but lies (a' ≠ 0), and only P occurs, he/she obtains the actual benefit (a), and pays (1 - p) (a + a'), which reports a net benefit of

$$pa - (1 - p)a' \quad \text{if P occurs}$$

It can be seen that the payoff decreases as a' increases.

On the other hand, if Q occurs, agent A will get back the amount (a + a') and pay (1 – p) (a + a'), which implies a net benefit of

$$p(a + a') \quad \text{if Q occurs}$$

This monetary amount increases with a'. Table 6 summarizes the benefits from the two events.

Table 6. Payoffs from events P and Q

Event	Value	Gross benefit	Premium cost	Net benefit
P	a	a	(1 - p)(a + a')	pa - (1 - p)a'
Q	0	(a + a')	(1 - p)(a + a')	pa + pa')

With "dishonest" insurance, agent A gets the same net mean expected return –as derived from (4) - than with an honest insurance. However, the difference in the net benefit between P and Q is precisely a'. Agent A fixes the variance he/she wants, since all the benefits but a' are regardless of the event that occurs. As a result, the larger the bias a', the larger the risk will be. And for an agent who does not subscribe to the insurance, the risk gets larger with his/her actual value a, thus giving a greater incentive to participate in the insurance market.

If an agent is risk averse, he/she will prefer the less risky option if both offer the same expected return but different risk levels. In this case, the preferred option is a' = 0. In other words, he/she will prefer to reveal an unbiased valuation.

Obviously, the case of agent A can be extended to the other agents, B, C, etc.... No agent will have an incentive to lie, if he/she is risk averse, which was the desired result for the mechanism proposed. Table 7 sums up the different results.

Table 7. Different insurance combinations

	Net benefit Without insurance	Net benefit with Insurance: Dishonest (a' ≠ 0)	Net benefit with Insurance: Honest (a' = 0)
Event P	a	pa - (1 - p)a'	pa
Event Q	0	pa + pa'	pa
Expected return	pa	Pa	pa
Benefit range	[a] - [0] = a	[pa - (1 - p)a'] - [pa + pa'] = a'	[pa] - [pa] = 0
Does risk exist?	YES	YES	NO
Risk increases	with value of a	with value of a'	NO

A remarkable characteristic of this mechanism is that the incentives to reveal true values are regardless of the probability p. However, the net benefit for consumers increases with p. On the other hand, the government (which issues the insurance) collects more money as p decreases. This would usually pose a problem, since the government would have incentives to collect more. In this case, though, the probability is observable (either there is a drought or there is not), which makes the problem less likely to happen.

The Supply Side

Once the actual demand is known, the question is to determine the optimal investment I* in desalination to increase the probability p of not having water shortages. In other words, the value I* that maximizes the net social benefit has to be determined.

Let p(I) be the function that relates the non-drought probability with the investment level, with $p'_I > 0$, $\forall I \geq 0$. Suppose that an investment of value I is undertaken during the initial time period (0). As a consequence, a value V (associated to event P with probability p) is obtained at each time period during T periods, as reflected in table 8.

Table 8. Investment and returns per each time period

0	1	...	T
- I	pV	...	pV

The Net Present Value (NPV) of the investment is

$$NPV(r) = -I + pV[1/(1+r) + 1/(1+r)^2 + ... + 1/(1+r)^T] = -I + pV[1/r - 1/r(1+r)^T], \quad (5)$$

which yields

$$NPV(r) = -I + pV/r \quad (6)$$

when T is large, with r being the discount rate.

The optimization program is

$$\text{Max } NPV = -I + pV/r$$

$$I \geq 0$$

and the marginal optimality condition is

$$p'V = r \quad (7)$$

The condition means that investment has to reach the level where its marginal return $(\delta p/\delta I)\cdot V$ equals the marginal cost of capital, r. Since it is not certain that the investment generates net benefits, the optimal level I* is desirable when NPV(I*) > 0, and there should be no investment whenever NPV(I*) ≤ 0.

For instance, let V' = 2000 be the marginal social value of securing a water supply, and let the relationship between the investment level (I) and the non-drought probability (p) be the one shown in table 9.

Table 9. Relationship between investment and probability of non-drought

I	0	500	600	800	1200	1800	2750	4000	7000	9000	15000
p	0.3	0.35	0.4	0.45	0.50	0.55	0.60	0.65	0.70	0.75	0.80

Applying the optimality rule – adapted to the example case, which is discrete - and taking an opportunity cost of capital of r = 10%, the following expression can be found:

$$(\Delta p/\Delta I)\cdot V = r = 10\ \% \qquad (8)$$

Table 10 shows that the optimal level of investment in the example would be I*=2750 monetary units. This would ensure a water supply with a chance of 60%, with a NPV of 9250. Note that this is larger than NPV(I = 0) = 6000. Therefore the ΔNPV = 3250 > ΔI = 2750.

Table 10. Optimal level of investment in relation to the other variables

I	0	500	600	800	1200	1800	2750	4000	7000	9000	15000
P	0.3	0.35	0.4	0.45	0.50	0.55	0.60	0.65	0.70	0.75	0.80
(Δp/ΔI)V	-	20	100	50	25	16.67	10.52	8	3.33	5	1.67
NPV	6000	6500	7400	8200	8800	9200	9250	9000	7000	6000	1000

For this market to work correctly, the insurance agency needs to have incentives to apply the optimality rule in (7). Suppose the insurance agency wants to maximize its benefits. In order to obtain an efficient result it would be sufficient for the agency to pay the quantity V in case of drought (event Q with probability 1-p), and receive an inflow of S (or its equivalent s per period, where s/r = S) that ensures its financial balance. The insurance agency program would then be

$$\max_{I \geq 0} NPV = S - I - (1 - p)V/r,$$

that is,

$$\max_{I \geq 0} NPV = S - I - V/r + pV/r \qquad (9)$$

Since both S and V/r are constants, (9) yields the same marginal optimality conditions, p'V = r, as the previous programs. It is therefore socially desirable to have the agency maximizing its net benefits.

Some Remarks on the Provision of Insurance of Water Availability

Before estimating the optimal amount of desalinated water to be provided, it is advisable to characterize the good from an economic perspective. It is especially important to ascertain whether – or to what degree - the good is consumptive. Depending on the good's characteristics, it is not reasonable to expect that the market will provide efficient results by itself. Both consumers and producers may have incentives (additional personal benefits) to behave strategically. As a consequence, it is advisable to design procedures that are compatible with individual incentives, and therefore individual decisions will lead to efficient results.

For this, a mechanism has been proposed for the demand side, based on an insurance subscription. In its original version, the mechanism only operates on the demand side, and it is aimed at choosing between two alternatives. The mechanism proposed here, however, works for the selection of the optimal provision of collective goods.

Furthermore, in the original mechanism the probability of the event is exogenous, unobservable, and plays no role in it. Furthermore, agents have to believe that the exogenous probability given is the actual one. In the adaptation carried out here, the probability is an endogenous variable, is observable, and is therefore less likely to be manipulated.

Since the supply side has also been modeled in a way that meets the incentive compatibility criteria for producers, the model is complete, and the expected results ought to be close to the optimum.

THE PROBLEM OF EXTERNALITIES

Characterizing the Problem and Refining the Concept

An externality is any cost or benefit that is imposed involuntarily and without compensation to any person. An externality arises when the range in which decisions are taken is smaller than the range in which the costs and benefits of an economic activity take place. In this way, not all the relevant costs and benefits are accounted for, but only those that are inside the range of the decision – *the privates* - obviating the rest – *the external.* As an example, if the range of the decision for a thermal power station project is the country, the costs that acid rain causes in the forests of other countries will tend to be obviated, because they are *external costs*.

When the activities of an economic agent, either producer or consumer, affect the opportunities for activities by another economic agent, and the party causing this effect does not pay for it, then this agent is causing an externality. The destructive use of air, water or forests, as well as cultural production and technological innovation, thus affect subsequent opportunities for using these resources, in both production and consumption. In particular, the decisions taken by present generations positively and negatively affect the opportunities for action of future generations.

When the externalities are *shiftables* –Bird (1987)- the first receptor of the externality can stop receiving its effects by transferring it to others. Some typical examples are the transport of residuals away from the area where they are generated and the change in the route of heavy transport. The simplest and most direct way of transferring an externality is transferring the activity that causes it: e.g. the location of highly pollutant industries in developing regions constitutes a widespread application of the opportunities for transferring an externality. The transfer of externalities is automatic in some special cases, as happens with a great deal of the atmospheric pollution generated in a city. If one tries to solve this kind of externalities, it is preferable, *ceteris paribus*, to act the closest possible solution to the producer of the externality to avoid the extra social cost of the successive transfers.

If the quantity of externality is the same for all the persons suffering from it and coincides with the total quantity produced, the externality is said to be inexhaustible – Bator (1958) - i.e., that it has the characteristics of a *Samuelsonian* collective good. The reduction of the ozone layer as a consequence of consumption and production activities causes an externality for each person that does not depend on the quantity of individuals, although all of them suffer the same quantity of this impact. Signs and symbols have this characteristic, and in general, so do all elements related with culture, from a principle of physics to a literary work, and of course, the climate of the planet. The externality would be *locally inexhaustible* if it fulfills the definition, but only within a locality, e.g. modification of the microclimate in an area due to deforestation. An interesting case is the called *network externality* – Katz and Shapiro (1985) - in which the impact of the externality not only falls with the quantity of consumers but it increases, meaning it has *hypercollective good* characteristics.

It is normal to consider the externality as a flow, i.e. the externality occurs during a certain time period and when it stops the original situation returns. This is true of light pollution and air congestion. In this case there is a *flow externality*. However, in other cases the effects of the externality do not disappear when the activity that has produced it ceases. In this case there is a *stock externality* - Doeleman (1988) - because its effect is cumulative, as is the case with DDT in the human organism, the disposal of mercury in the sea and the production of radioactive waste. An activity that can cause a stock externality would not be relevant if the system is capable of absorbing the externality without an appreciable change in the basic characteristics of the affected resources. It would be then equivalent to a flow externality. However, the same activity emits a relevant stock externality if the resource crosses a certain threshold of relative scarcity that prevents assimilation without environmental cost. Again, there is a conflict between present and future generations - when a resource is extremely abundant for present generations it is logical that it is exploited at a rate at which it ends up being scarce for future generations. In extreme cases, a stock externality might cause irreversible situations such as the extinction of species which limit the opportunities of future generations. The current level of knowledge about natural processes does not always certain determination of whether an externality behaves as a flow or as a stock.

As is the case with goods, some externalities are transformable, which increases the possibilities of regulation and complicates the process for internalizing these costs. In some cases, it is possible to reduce the cost of an externality by means of simple transformations, such as the compacting of solid urban residuals. Furthermore, if a solid or liquid pollutant is burnt, it becomes another pollutant, which is maybe even more harmful and is more difficult to control as it is gaseous. This might benefit the producer, but it increases the inefficiency of the system as a whole.

In any case where there are negative (positive) externalities as a consequence of the production or consumption of a good, the market allocation of this good will be inefficient due to excess (shortage) as a result of the market not considering all costs and benefits - internal costs are computed while external costs are ignored. Solving a problem of externalities does therefore not mean eliminating the externality but only taking into account its entire value, that i.e. achieving the *internalization* of the external costs (benefits) in such a way that the consideration of the costs and benefits is complete.

An externality is said to be relevant in allocative terms if the social marginal benefit differs from the private marginal benefit, as the latter does not take into account the corresponding external benefit and/or when the social marginal cost is different from the private, as the latter does not incorporate the external cost. A typical example of an external marginal benefit would be the benefit for the whole society from an increase in the consumption of culture by an individual, and a typical case of external marginal cost is the pollution produced by cars.

It might be the case there was an externality in absolute terms, but that the marginal externality was zero in the optimum. In this case there is said to be an *inframarginal externality* – Buchanan and Craig (1962) - which is not relevant in allocative terms providing that the benefit provided by the externality exceeds the costs of avoiding it. Let us assume that there are two technologies for producing output X, A and B. The net private benefit of the activity produced with technology A is of $100, the total external cost due to pollution is $25 and the external marginal cost is zero in the point of optimal production. As the externality is inframarginal, the quantity decided by producers coincides with the optimal quantity and there are no reasons of efficiency to justify public intervention. However, with technology B, the private marginal cost would increase to $4 although the total external costs would be reduced to $1. The producer does not have incentives for adopting the new technology because his/her profits would be reduced, and as technology A does not generate marginal externalities there is no problem by hypothesis, which is absurd. There is no doubt that it is necessary to take all the externalities into consideration, and not only the marginal ones, in order to mitigate environmental problems.

The Solutions of Economic Theory to the Problem of Externalities

All the imaginable theoretical solutions to the problem of externalities are variants of a unique solution: *reassigning the property rights*. From an economic perspective, if there is no regulation of any kind, it is as if all the property rights - including the right to destruction -[3] were assigned to polluters. In fact, in the absence of regulation, any potential polluter can pollute by production or consumption activities with no need to pay for it in any way, with no need to ask for permission or give explanations to anybody, as if he/she possessed the entire property rights to the resource that is being polluted. If the cause of the problem lies in the actual allocation of the property rights to the polluters, it is therefore clear that any solution should take a reallocation of property rights into account. A reduction in the quantity of the property rights controlled by the polluters is required.

The basic ways of redistribution or reallocation of property rights are: a- Enlargement of the range considered. b.- Taxation a la Pigou, c- Administrative regulation, d.- Market

[3] They also enjoy the right of negotiation of these property rights, although it would not be used. On the one hand, as any potential polluter he/she has the same right, it would be nonsense that a potential polluter paid for something that could be freely obtained. In addition, the purchase of property rights in order to reduce pollution is problematic, because it would be necessary to buy the property rights of all possible polluters.

negotiation and e- Cultural regulation. As will be seen below, these solutions are relatively effective for solving the problems of present generations but do not allow the externalities affecting future generations to be solved.

a. Enlargement of the range considered. An externality is defined as all the impacts outside the range in which all costs and benefits are considered. Enlarging the range is thus enough to make the externality disappear. For example, a country that is part of the European Union has to accept EU discipline which includes the internalization of some externalities.
b. The taxation solution a la Pigou – the *polluter pays principle* - consists of a tax per unit of production or of consumption equal to the value of the external marginal cost, evaluated at the optimal quantity in the sense of Pareto. As it is a tax per unit, the marginal private cost increases by the exact value of the tax, the value of the marginal external cost, so the externality is internalized. Other taxes, such as the one on sales or value added tax, distort the relative prices of goods and cause inefficiency (*dead weight*). However, the tax that corrects externalities improves the efficiency of the system. For this reason, if the collecting capacity of the *Pigovioan* tax is used for reducing or eliminating a classical tax, then there is a double gain, known as a *double dividend.*
c. With the solution by means of an administrative regulation, the property rights pass to the state, which concedes pollution permits to some agents, in set maximum quantities and under certain conditions. Administrative regulation takes multiple forms and is applied in a global way, either specifically or combined. For example, the transit of cars in a city can be restricted when the total pollution overcomes a preset value[4]. This is an ideal solution if the prohibition of an activity (making fires in forests) or making it compulsory (primary education) is a good approximation to the optimum, because the administrative and compliance costs are minimal.
d. The market negotiation solution, the so-called *Coasian* solution, is based on free negotiation between two types of agents, those incurring in an opportunity cost if they cannot increase pollution and those that obtain greater utility as pollution falls. The state possesses all the property rights, transfers a part to some agents according to some prefixed criteria and organizes a market for those property rights. Negotiation takes place between agents that, given the market price of the property rights, received property rights in excess (supply) and those that wish to have a greater quantity of rights (demand), for polluting more or for preventing others from polluting. The resultant prices and quantities coincide with the *Pigouvian* solution under ideal conditions[5] and, they would be the same if all the property rights were given to polluters and if they were given to the victims. Under less ideal conditions, the result will not be efficient and the bias will depend on the distribution of property rights. With the property rights in the hand of polluters (victims), the inefficiency would always be an excess (shortage) of pollution.

---

[4] Paradoxically, the limit is not determined in relation to the total quantity of pollution produced but tc the pollution remaining after the winds have transferred a part to other areas.

[5] That is to say, under conditions perfect competition, without marginal transaction costs and with a negligible income effect.

Although in some cases it could be a useful solution for solving the externalities in a certain (developed) country, the consequences of applying a similar process between two countries – or two regions in the same country - can be regrettable if the difference in wealth between suppliers and demanders is a large one. In fact, there are many examples of exports of residuals from rich to poor countries at very low prices and with grave global consequences.

e. Cultural regulation, unlike the other solutions, only allows some kind of externalities to be solved - those that depend on individual agents, typically the consumers in some way. The reallocation of property rights is done in favor of society as a whole and there are social sanctions for those that do not follow the rules that are generally conveyed orally. This solution is appropriate as an alternative to administrative regulation and also as a complement to it. It is also useful for avoiding what Kant called the *tyranny of small decisions*, which consists of a high social cost (benefit) caused by multiple individual decisions, each implying a negligible cost (benefit). The main advantage of this kind of solution is its relatively low cost and its stability over time.

Limitations of the Solutions to Externalities

In spite of the quantity and variety of good instruments for internalizing externalities, the problem should not be expected to be easily or quickly solved. The internalization of main externalities has a cost, which judging by the policies followed by most countries – hesitant and slow - is too high to be assumed by the current economic system. It seems that the participation of various generations will be necessary in order to assume the economic costs of taking environmental externalities into account.

An important restriction is the one derived from the current administrative structure available. With the exception of countries with a powerful and efficient administrative organization, which is able to detect any negative deviation and solve it immediately, none of the classical solutions could be applied in the form and intensity needed - even the *Coasian* market solution.

The limitations that are inherent in each kind of solution manifest themselves in real applications. Any way of internalization - the *Pigouvian* way or any other - necessarily affects costs and so affects market prices. This is at odds with the need to be competitive in international markets, because countries that do not internalize externalities can offer the same product at a lower price (*green dumping*). Any solution to the problem is therefore more easily applicable in local than in international markets.

In the regulation of externalities, some of the agents causing them are usually unaffected by the correcting tools. For example, if an urban toll is established to reduce city congestion and pollution, it is highly probable that those using ambulances, police vehicles and fire engines, among others, will not be not subject to this tax. Logically, the toll tries to reduce journeys in private vehicles but, undoubtedly, it does not try to stop the activity of the various public services. However, this fact complicates the computation, because now the activity of the users that are exempt from the tax should be subtracted from the optimal quantity in order to determine the remaining quantity for the non-exempt users. Under these conditions, the first best solution would not be achieved, but instead the second best solution in the best of cases.

In any case, the most serious limitations appear when trying to solve intergenerational externalities. Intergenerational externalities exist because the actions of present generations affect the possibilities of future generations. There is also a kind of externality consisting of the fact that present generations have defined preferences concerning the consumption or welfare of their descendants, future generations. There are therefore externalities between present generations and future generations that should be internalized, either if the objective is a redistributive improvement or if it is merely an increase in economic efficiency.

It should be remembered that any imaginable solution to an externalities problem requires an expropriation of propriety rights because in the absence of regulation, these belong completely to the agents that cause them, i.e. the present generations in this case. In the case of the externalities affecting future generations, if these external effects have to be internalized, at least a part of the property rights should therefore be transferred from present generations to future generations. For the moment, there is no practical way of carrying out this transfer and all the property rights belong *de facto* to present generations. The means of management of these natural and environmental resources is therefore *free access* for the present generations, which is the most inefficient of the possibilities.

The limitations of classical solutions are very serious in the intergenerational context. In the *Pigouvian* solution, for example, it is necessary to determine the value of the external marginal cost. In the case of a non-durable externality, such as noise, it is "enough" to find out the valuation of present generations. However, when the externality has durable effects, as is the case when nuclear waste is generated, a language is lost, a species disappears, or pollutants are continuously dumped in the sea, the computation of this cost is different if one considers the perspective of future generations than if the interests of future generations are also taken into account.

The application of a solution *a la Coase* has serious problems. As future generations cannot act in any way in the property rights market, because they do not exist yet, and because they do not have a representative that can negotiate in their name on equal terms with other agents, this solution is not applicable.

It is unreasonable to expect an administrative regulation by means of norms for internalizing externalities to appropriately take into account the interests of future generations, given that future generations do not have anyone representing them before those design and apply the norms.

The same will occur with a *Pigouvian* solution with an externality only affecting future generations. The government does not have incentives to burden its potential voters with a tax benefiting of their descendants, because future generations do not have either a voice or a role in the political arena.

Furthermore, if a member of present generations sees that his/her interests have been harmed, he/she has recourse to the administrative and law system to defend his/her interests. However, this is not possible for future generations, so the possible harm that they suffer would go unpunished, in as far as that they do not have a legally recognized representative that could claim damages in their name.

In conclusion, as any solution needs from the reallocation of property rights, and as future generations do not have any system which allows them to defend their property rights, it can be stated that, for the moment, economic theory does not provide any mechanism for internalizing the externalities that would affect future generations.

An Application of a Case of Externalities

The Choice of the Best Use for a Space

There is a limited space available that can be used in only way from use among different incompatible alternatives. The gross social value of the space in its current state is V^0, and the value corresponding to the different alternatives is $V^1, ..., V^n$, respectively. Adapting the space to use j involves a transformation cost $C_j \geq 0, j = 1, , ..., n$. Initially, two cases can happen:

Case 1.

$V^0 > V^j, \forall j$. In this case, there is a sufficient condition for the transformation of the use of the space, which we call project T, not to be interesting under any circumstances and so the problem is already solved. The best option is the current one, the 0, and the value of the space is V^0.

Case 2.

$\exists h$ such that $V^0 < V^h$, $h = 1, ..., H$. If there is at least one use providing a greater social value than the current use, then there is a necessary condition for project T to be desirable, but this condition is not a sufficient condition. Two things might happen, either $V^0 > V^h - C_h, \forall h$ or $\exists k\ V^0 < V^k - C_k$, which is examined next.

a. $V^0 > V^h - C_h, \forall h$. Under these circumstances, the project T is not desirable because the condition is necessary and sufficient for the non-desirability of modifying the current use. In spite of the fact that the use 0 is not the best, because there are better alternatives, the cost of transformation is high enough to prevent the transformation. The value of the space is still V^0, both *ex ante* and *ex post*. If, for any reason, the space was transformed by giving it another use, e.g. use s, then the value of the space *ex ante* would be equal to the maximum value among V^s and $(V^0 - C_s)$ and the value *ex post* would be of V^s.
b. $\exists k\ V^0 < V^k - C_k$, $k = 1, ..., m, m+1, ..., K$. The current use 0 is not optimal because there are K better alternatives and, besides, the change of use implies an increase in value which more than compensates for the cost of transformation C_k. Here there is a necessary and sufficient condition for the desirability of the transformation of T.

It could be the case that V^0 was taken as the valuation *ex ante* of this space, but this, as we will see, would not be correct at all. Before starting the valuation task, the soundness of the K alternatives should be examined. The best alternative in relation to the *status quo* is the use m that fulfills $V^m - C_m \geq \max\{V^k - C_k\}$, and this is the best use of the space[6]. As a hypothesis, we have $V^0 < V^m - C_m$, the maximum that one would be willing to pay for the space in the state 0 is $V^m - C_m$, and the value *ex ante* of the space is for this reason, $(V^m - C_m)$.

[6] Under ideal conditions, as a perfect capital market. In other cases -for example, if there were financial restrictions-, there is the possibility that another alternative would be preferable.

Carrying out the project T, the *ex post* value changes to V^m, because the cost C_m is a *sunk cost* as it cannot be modified in any way and thus, it should not be taken into account.

To summarize the above,

Cases	0 is optimal	Best use		T, desirable	Type of condition	Space value	
						ex ante	*ex post*
$V^0 > V^j$	Yes	0	⇒	No	sufficient	V^0	V^0
$V^0 < V^h$	no	h	⇐	Yes	necessary		
$V^0 > V^h - C_h$	no	0	⇔	No	necessary and sufficient	V^0	V^0
$V^0 < V^m - C_m$	no	m	⇔	Yes	necessary and sufficient	$V^m - C_m$	V^m

The Distribution of a Space Among Two Incompatible Uses

Let us assume that there is a total quantity of space $\underline{e}$, which cannot be increased in any way. There are two alternative uses of this space — α and β, which are incompatible. The problem is that of separating those uses and distributing the total space between both uses in an optimal way.

Let us assume there are a users of type α and b users of type β, and that all the users of the same type are identical. The users j have defined preferences over the goods $X_{j,\ j\ =\ \alpha,\ \beta}$, and Y. Y is a pure private good that is available in a predetermined quantity w. The goods X_j are local collective goods, which are produced[7] using the private good as an input (Z_j) according to the functions

$$X_j - f^j(Z_j) = 0 \quad \phi_j \tag{10}$$

being ϕ_j the multipliers associated to shadow prices constraints. As X_j are local collective goods, all the users j will consume the total quantity produced X_j.

The total quantity of private good *w* is distributed between its use as an input and its use as a consumption good,

$$\underline{w} - Z_\alpha - Z_\beta - aY_\alpha - bY_\beta \geq 0 \quad \lambda \tag{11}$$

with λ being the shadow price of the private good.

Consumption of the good X_j requires the space e_j. The total quantity of available space is prefixed in *e*, so it is not possible to modify it

$$\underline{e} - e_\alpha - e_\beta \geq 0 \tag{12}$$

with σ being the shadow price of the space as a resource.

The benefit obtained depends on the quantity of goods X and Y consumed, which constitutes a gross social benefit (SB) of

---

[7] Instead of interpreting $X_j = f(Z_j)$ as the result of a production process it can be seen as the access cost to good j that provides the space e_j.

$$SB = a\prod^{\alpha}(X_{\alpha}, Y_{\alpha}) + b\prod^{\beta}(X_{\beta}, Y_{\beta}) \quad (13)$$

The two uses of the space generate negative externalities that are locally inexhaustible. The costs (SC) of X_j depend on the quantity Z_j of the private good employed in its production, on the number of users on the activity j, and on the quantity of available space for each use j,

$$SC = C^{\alpha}(X_{\alpha}, e_{\alpha}, a) + C^{\beta}(X_{\beta}, e_{\beta}, b) \quad (14)$$

The objective is to maximize the net social benefit, NSB = SB – SC, taking into account the production function of X_j and the constraints on the quantity of space and private good Y. The control variables of the maximization program are the quantity of space allocated to each use (e_{α}, e_{β},), the quantity of users (a, b) of each type (α, β), the quantities of the collective goods and the distribution of the private good. Both the quantity of total space *e* and the total income *w,* are problem data and cannot be modified.

The marginal conditions of optimality are quite simple:

$$\partial\prod^{\alpha}/\partial Y_{\alpha} = \partial\prod^{\beta}/\partial Y_{\beta} \quad (15$$

$$\partial C^{\alpha}/\partial e_{\alpha} = \partial C^{\beta}/\partial e_{\beta} \quad (16)$$

$$\partial f^{j}/\partial Z_j + \partial C^{j}/\partial X_j = i\partial\prod^{j}/\partial X_j, j = \alpha, \beta, i = a, b \quad (17)$$

and the interpretation is quite direct:

a. To distribute the quantity of the private good in such a way that there is equality in the marginal benefit for all users
b. To distribute the space between the uses j in such a way that the marginal benefit derived from the space is the same for both spaces
c. To produce the collective goods j in such a way that the sum of the marginal benefits of the users of each type is equal to the cost of production – the *Samuelsionian* condition for local collective goods - plus the cost of the marginal externality – the *Pigouvian* condition for internalizing externalities.

Some Remarks on the Problems of Externalities

The set of economic instruments for internalizing externalities and providing collective goods is more or less satisfactorily effective in solving the environmental problems affecting present generations. However, these same instruments are inappropriate for considering the externalities that present generations inflict on future generations and for preserving intergenerational goods.

If the world ended with present generations, problems such as climate change or the loss of biodiversity would merit much less attention. The reason for this is simple - the costs for present generations are relatively low, while they are expected to be very high for future generations.

It should be remembered that a necessary condition for the internalization of intergenerational externalities is the existence of a system which allows future generations to have a part of the property rights and to manage and negotiate them with no limitations other than those governing the behavior of present generations. At present, there is no institution which allows property rights to be transferred to future generations. Efforts towards environmental management following conventional economic theory therefore only work in favor of the interests of present generations, in as far as they obviate the interests of future generations.

We understand that the main problem we are faced with, is the conflict between the interests of present generations and future generations. It is for this reason that we devote the second part of this work to analyzing the consequences of this conflict, from both a distributive and from an allocative perspective, focusing our analysis on the appraisal of intergenerational policies and projects, which are usually used to help social decisions in environmental problems.

Future Generations in Project and Policy Appraisal

Conventional evaluation and management methods, such as the net present value, have been widely criticized because of their discrimination against the interests of future generations. These methods do not consider the existence of different generations and thus ignore their interests. In short, their applicability to problems with strong long-term effects has been questioned. This is the case of problems like climate change, the ozone layer hole, deforestation and the loss of biodiversity. In this section, we analyze the limitations of conventional economic evaluation of intergenerational problems and propose an alternative for application to projects affecting future generations.[8]

The Limitations of Conventional Evaluation of Long Term Projects: Discounting and Future Generations

Conventional economic analysis gives less importance to flows that take place in the future (a thorough review of the problems of conventional discounting can be found in Broome (1992) and Price (1993, 1996)). The application of conventional time discounting devalues and practically removes the impacts that occur in the distant future from the analysis. A higher discount rate implies greater discrimination against future generations, although any positive discount rate leads the analysis to strongly devalue and almost ignore distant impacts.

The social discount rate (s) usually used to discount future impacts is expressed with the Ramsey formula:

$$s = \rho + \eta g$$

[8] See Padilla (2002) for a broader analysis of these limitations and a previous version of our proposal.

where ρ is the pure time preference rate, η is the elasticity of marginal utility (absolute value) of consumption and g is the growth rate of per capita consumption, i.e. discounting is applied because of impatience and the belief that there will be more wealth per head in the future.

Conventional cost-benefit analysis applies the time discount of the present society to discounting all costs and benefits that will occur in the future, as if all future impacts happened to present individuals. The consumption of future individuals is discounted using a rate that shows the impatience of present society, while the logical procedure would be to consider the preferences of present society for the well-being of future generations. An intergenerational weighting appropriately showing these preferences should be applied.

The practical application of the argument of decreasing marginal utility of consumption is also controversial. Applying a high discount rate because of assumed future prosperity could lead to compromising this very prosperity by giving low weight to future impacts (the 'optimist paradox').

The individuals originating long-term problems and the ones suffering their consequences are certainly not the same individuals, and we are not certain that these future individuals will enjoy greater well-being. Neither the argument of decreasing marginal utility nor the pure time preference therefore justifies the application of a constant time discounting to long-term impacts.

Many economists still use a much cruder basis for discounting, i.e. the opportunity cost of investment funds. In this way, the resources would be placed in the highest yielding projects, thereby obtaining greater future well-being. This argument is hardly justifiable in the intergenerational context, as it is based on the full reinvestment of the revenues obtained from the exploitation of the resources, which has not occurred in the past and is unlikely to occur in the future (Price, 1996).

The weights to apply to future generations should reflect the altruistic preferences of society and not just be an arbitrary extension of the time preferences of the present (this is analyzed in Section 5). However, considering these altruistic preferences does not guarantee that the interests of future generations are appropriately taken into account. If future generations have certain rights that should be respected, these rights should be included in the analysis.

Intergenerational Equity and the Sustainability Requirement

Conventional economic analysis implicitly assumes that the Earth and all its resources belong exclusively to present individuals. What is more, the present has the power to decide how to use these resources. As a consequence, the endowment that will reach the hands of future generations is just a residue of the decisions of present individuals and not the result of a negotiation or market including the interested parties. Although there are altruistic preferences that should be considered, these do not solve the entire problem. For a satisfactory solution, both parties (present and future) should arrive at an agreement (which is certainly not possible) or make the recognition of certain rights to future generations based on moral grounds explicit, and act accordingly.

Strong moral, contractual, or deontological justifications lead to rejection of the premise 'everything belongs to the present' as a legitimate starting point. Conventional analysis takes the market as the only relevant condition, arguing that this procedure is not value laden.

However, this option implies both denying any right to the unborn since they cannot participate in present markets, and accepting that the present can do whatever it pleases without any limit, which certainly is strongly value laden. In order to respect the interests of future generations, present actions should be kept within some limits. It would be more appropriate if these were subject to moral restrictions or, if as a result of a contractual vision, assumed sustainable development as an implicit moral agreement between generations (Barret, 1996; Howarth, 1997).

Much of the literature on sustainability tries to establish different criteria compatible with a development that can be sustained over a long course of time, i.e. compatible with a minimum of intergenerational equity. The most popular definition states that sustainable development is the "...development that meets the needs of the present without compromising the ability of future generations to meet their own needs." (WCED, 1987, p. 8). This definition implies limiting the present use of resources.

From an optimistic point of view, if every time that natural capital diminishes, an increase in the equivalent manmade capital takes place, then the capacity to maintain the quality of life would not be affected. This is known as the 'weak sustainability' criterion. It somewhat further than conventional economic tools such as cost-benefit analysis because it denies the validity of the Hicks-Kaldor test in the intergenerational context and demands actual compensation if future generations suffer from present actions (Neumayer, 1999). However, it is based on the unreal assumption of perfect substitution between different types of capital. Sustainable development requires making the limits to substitution part of the analysis, and not making very uncertain assumptions that might jeopardize future generations. The respect for future generations' interests needs the implementation of the 'strong sustainability' requirement, a criterion that requires the maintenance of the quantity and quality of natural capital over time, and of some critical levels of certain types of natural capital.[9] Costanza (1994, p. 394) justifies it as a "...prudent minimum condition for assuring sustainability". Ignorance and uncertainties about substitution possibilities imply that there is no feasible way of putting the weak sustainability criterion into practice on a global scale in order to assure sustainability. Present generations will ensure fair treatment for future generations if the levels of the different types of capital do not diminish.

The sustainable development requirement implies assuming a much more favorable distribution of rights for future generations than that contained in conventional economic analysis. Page (1983) states that the life opportunities of future generations will be undiminished if they inherit the same resource base as that inherited by present generations. Bromley (1989) analyzes the possibility of undiminished stocks of natural resources and environmental quality for ensuring intergenerational justice. Finally, Howarth (1997) asserts that the provision of specific endowments of reproduced capital, technological capacity, natural resources and environmental quality may sustain life opportunities for future generations. In conclusion, in order to guarantee fair treatment for future generations, the analysis should consider their right to an undiminished socioeconomic and ecological capacity. The capacity of the present to alter the conditions of life of the future imposes this responsibility.

---

[9] Its quantification might be quite problematic; but it is in any case possible to have some indicators of its evolution.

The recognition of rights to the future, with the application in each case of the necessary proceedings, requires new evaluation and management methods integrating the concerns expressed throughout this work as well as strong institutional support reinforcing it.

Evaluation Incorporating the Sustainability Requirement and Intergenerational Weighting

This section proposes the application of a new evaluation process incorporating the sustainability requirement and intergenerational weighting. The recognition of rights implied by the sustainability requirement leads to the analysis of intergenerational problems incorporating the obligation of maintaining the economic and ecological capacity that is currently enjoyed. This requires a different evaluation of policies and projects, depending on whether the structure of rights between generations is affected (intergenerational evaluation) or not (ordinary evaluation).[10]

Ordinary Evaluation

In the actions that only affect the generations taking the decision, conventional evaluation and management methods are essential for determining the most efficient allocation of resources. Nevertheless, there can be important distributive, ecological, moral, affective, contractual, deontological, cultural, or political reasons for society. These should also be considered in decision-making. As standard cost-benefit analysis follows only the allocative efficiency criterion, in some cases it could be appropriate to include it as merely an element of a more complete decision-making process.

Intergenerational Evaluation

When a project has effects on generations not participating in the decision process, various cases should be differentiated. There are projects that will not negatively affect future generations, but there are also some that might harm them. This separation cannot be made with economic information alone - the need to consider geophysical and ecological conditions makes the evaluation and management of sustainability into an interdisciplinary task.

Intergenerational projects not jeopardizing the capacity of the future, in principle, do not give rise to a transaction of rights between generations and therefore do not imply obligations for the present. However, the fact that present generations are (implicitly) assumed to enjoy any future consumption shows that conventional valuation is quite arbitrary. Following the reasoning of subsection 4.1, an intergenerational weighting should be applied that properly shows the preferences of current society regarding the consumption of future generations.

Any action jeopardizing the opportunities to be enjoyed by future generations implies a transaction of rights between generations. In each case, it will be necessary to consider which way of fulfilling the obligations to the future is least costly to the present. The following relevant options could be considered:

a. *Not to carry out the project:* If a project causes irreversible harmful effects to future generations and these cannot be avoided or compensated for, it should be considered

[10] Page (1997) arrives at similar conclusions and differentiates between ordinary and constitutional decisions while Norton (1995) proposes a two-tier interdisciplinary approach.

as being outside the choice of possibilities. This is the case with exploiting renewable resources indefinitely beyond their regeneration rate, or overexploiting the assimilative capacity of the environment. In addition, when there are serious risks or uncertainties, obligations to the future imply a bigger risk aversion in decision-making. Basic processes and some critical levels necessary for the sustenance of the ecological system should be protected. The information of different scientific disciplines should help to determine which goods require this protection.

b. *To undertake precautionary and control measures:* If the modification of the structure of rights that the original project implies is avoidable (e.g. enhancing security systems) and it is still profitable, this option is more appropriate than the first. Conventional computation of costs and benefits (ignoring the future) often leads to ignoring the adoption of security measures or clean technologies, even if they could prevent severe harm to future generations. The obligations of present generations imply that these measures should be included within the unavoidable costs of the project.

c. *Compensation through an associated project:* In some projects it is possible to compensate for the harmful effects on future generations through an associated project (e.g. reforestation).[11] The cost of the compensation should be included in the calculation of profitability, and the way in which this became effective should be articulated. The rights of future generations would thus be protected, with an exchange taking place between generations. In order to permit this transaction of rights a sine qua non requirement should be that decision-makers demonstrate that this compensation will be sufficient and will become effective.

d. *Financial compensation:* This option would clearly modify the composition of the capacity bequeathed to future generations. There should not be any doubts about the possibility of substituting the diminished resources and of establishing an investment fund enabling this future compensation.[12] The damage caused to future generations must therefore be quantifiable in monetary units or, at least, it must be possible to demonstrate that the compensation will, in all likelihood, be satisfactory. Once again, and unlike conventional methods, it should become effective.

The evaluation process, should determine in each case which option is more appropriate and, as argued above, economic information is not enough to decide this. Figure 1, taken from Padilla (2002), outlines the evaluation process proposal.

---

[11] This option coincides with the proposal of Markandya and Pearce (1988) on shadow projects.

[12] Costanza and Perrings (1990) idea of environmental bonds could be seen as a way of coming into operation this compensation.

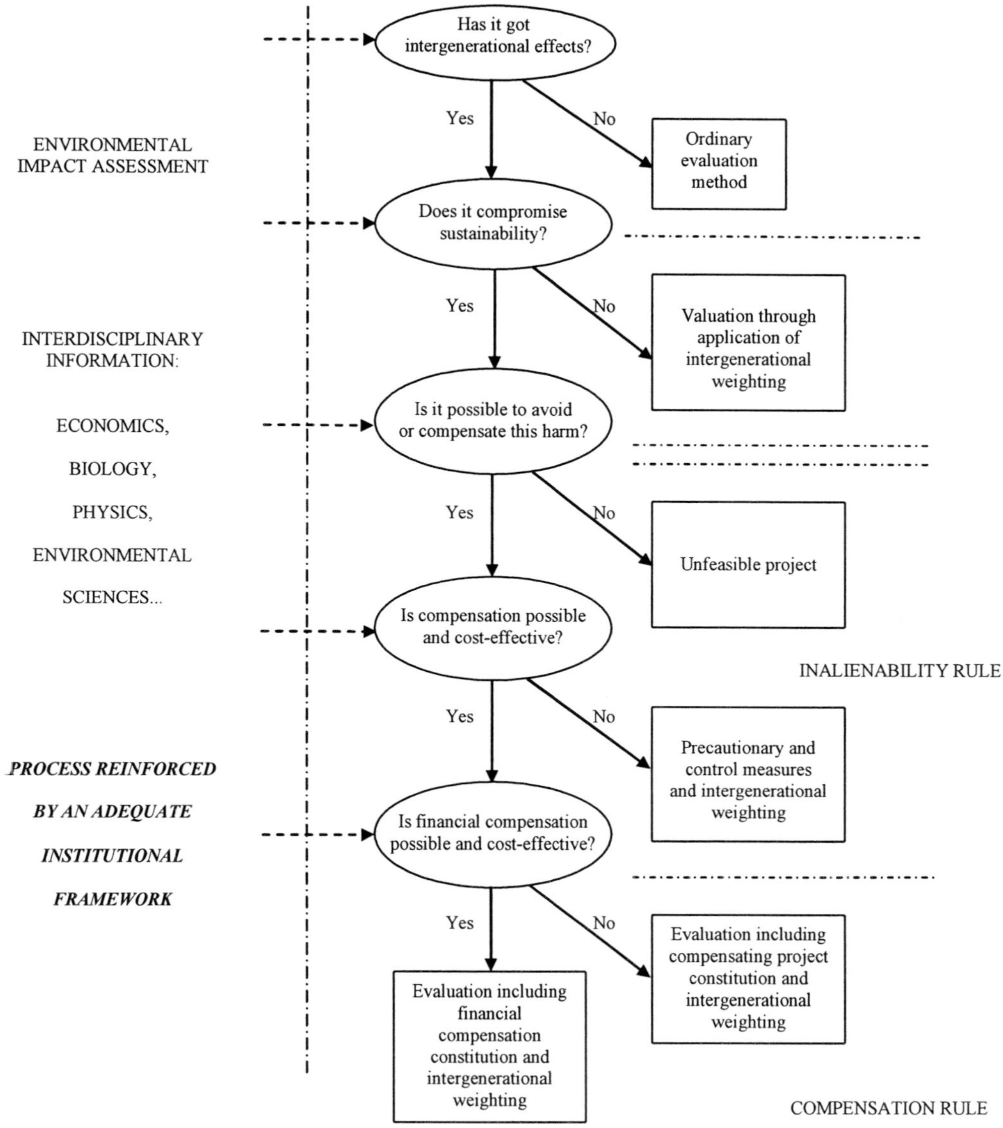

Figure 1. The evaluation process

Intergenerational Weighting in Long-Term Project and Policy Appraisals: The Multigenerational Net Present Value

In the previous section we have highlighted that conventional evaluation methods have been broadly criticized because they disregard the impacts occurring to future generations. Their application of time discounting gives a negligible weight to future generations, besides not being very fair, it does not accord with the preferences of society. Applying the time preference of present generations beyond their life-span, as if they were immortal, ignores the fact that these impacts will occur to other individuals. This section tries to overcome this arbitrary premise of conventional economic analysis and studies the consideration of the impacts happening to future generations through an intergenerational weighting which appropriately reflects the preferences of society.

Intergenerational Weighting in A Successive Generations Model with Intergenerational Altruism

First, the paper considers an economy with successive (non-overlapping) generations. In this model, society is composed of one generation of N individuals, which at the end of each period is substituted by another generation of N individuals (their descendants). The utility of present generation individuals is influenced by the utility level of their descendants (non-paternalistic altruism), so the life utility of an individual is:

$$U^n_t = V^n_t + \beta U^n_{t+1} \qquad (18)$$

Where V^n_t is the life utility function derived from own consumption of individual n of generation t; $\beta=(1+R)^{-1}\in(0,1)$ is the altruistic parameter; U^n_{t+1} is the life utility function of the descendant, individual n of the next generation.

The utility function can be expressed as the next weighted sum of the satisfaction that the future descendants derive from consumption (V^n_g):

$$U^n_t = \sum_{g=t}^{\infty} \beta^{g-t} V^n_g \qquad (19)$$

So, the intergenerational weighting that an individual of generation t applies to his/her descendant of generation g is β^{g-t}.

One of the principal premises of cost-benefit analysis and welfare economics is that the preferences of the individuals that constitute a society should be taken into account. Following the utilitarian criterion, the social welfare function is expressed here as the sum of the utilities of the individuals that comprise society:

$$W = \sum_{n=1}^{N} U^n_t = \sum_{n=1}^{N} \sum_{g=t}^{\infty} \beta^{g-t} V^n_g \qquad (20)$$

Hence, the social intergenerational weight applied in moment t to individuals of generation g is $\beta^{g-t}=(1+R)^{t-g}$, which is equal to the individual weight shown in (19). However, this direct result does not hold if the coexistence in society of more than one generation is considered, as shown in the next subsection.

Intergenerational Weighting in an Overlapping Generations Model with Intergenerational Altruism

Overlapping generations models, introduced in economic analysis by Samuelson (1958), provide the necessary tools for dealing with the coexistence of different generations in society. In the model considered in this section, society is composed of two generations of N individuals each. Thus, in moment t there are N old individuals of generation t-1, with the next utility function:

$$U^n_{t-1} = \sum_{g=t-1}^{\infty} \beta^{g-(t-1)} V^n_g \tag{21}$$

And there are N young individuals, of generation t, with the next utility function:

$$U^n_t = \sum_{g=t}^{\infty} \beta^{g-t} V^n_g \tag{22}$$

Therefore, the intergenerational weighting that any individual of future generation t applies to his/her descendant of generation g is β^{g-t}, which is higher than $\beta^{g-(t-1)}$ the one applied by the ascendant of generation t-1.

Society is composed of two generations that weight the consumption of future generations differently. The valuation of own consumption of an individual of generation t+1 influences the utility of his/her parent (present generation t) through weight β, and in the utility of his/her grandparent (present generation t-1) through a smaller weight, β^2. In the same way, the remaining future generations influence to a different degree the utility of different present individuals according to their propinquity. The social welfare function in moment t is:

$$W = \sum_{n=1}^{N} U^n_{t-1} + \sum_{n=1}^{N} U^n_t \tag{23}$$

Which can be expressed as follows:

$$W = \sum_{n=1}^{N} V^n_{t-1} + \sum_{n=1}^{N} \sum_{g=t}^{\infty} (\beta^{g-(t-1)} + \beta^{g-t}) V^n_g \tag{24}$$

Thus, the social intergenerational weight (μ_g) to apply in moment t to any individual of generation g > t is:

$$\mu_g = \beta^{g-(t-1)} + \beta^{g-t} = (1+R)^{-(g-(t-1))} + (1+R)^{-(g-t)} \tag{25}$$

Clearly, the social weighting to the consumption of future generations derived from social preferences is not equal to any individual weighting. In addition, it clearly differs from the arbitrary weighting that conventional discounting involves:

$$(1+r)^{-(g-t)} \neq (1+R)^{-(g-(t-1))} + (1+R)^{-(g-t)} \tag{26}$$

Inequality (26) shows that in moment t the weight discounting applies to generation g (left hand side) notably diverges from the weighting derived from social preferences (right-hand side). Moreover, it is not possible to express the social weighting as a simple discount factor.

The Multigenerational Net Present Value

As shown previously, conventional discounting applies an arbitrary weight to future generations. Some authors (Collard, 1981; Nijkamp and Rouwendal, 1988; Bellinger, 1991; Padilla, 2002; Pasqual and Souto, 2003) have advocated in favor of substituting the weights of the time discount factor for an intergenerational weighting applying an explicit weight to each generation. However, they do not specify much about how these weights should be determined. This paper's view is that intergenerational weighting should show the preferences of society regarding the consumption of future individuals.

There is a broad acceptance that the net present value (NPV) is an appropriate method for calculating the profitability in projects affecting only one generation or one individual. Considering that each generation discounts its consumption flows through its own time discount, NPV_t might be defined as the value that individuals of generation t assign to the increase in own consumption generated by the project. We propose the use of the next weighted sum, the multigenerational net present value (MGNPV), for obtaining the social valuation of the project:

$$MGNPV = NPV_{-1} + \mu_0 NPV_0 + \mu_1 NPV_1 + \ldots + \mu_T NPV_T \tag{27}$$

Where T is the last generation affected by the project.

The weight to apply to each generation (μ_g) should correspond to the altruistic preferences of individuals that compose society. In the model of two overlapping generations, $\mu_g = (1+\beta)\beta^g$, (except for μ_{-1}, which is 1), and the derived social valuation is:

$$MGNPV = NPV_{-1} + \sum_{t=0}^{T} (1+\beta)\beta^t NPV_t \tag{28}$$

This method might avoid the negligible weight conventional NPV applies to future generations in long-term projects. Moreover, it takes into account the existence of different generations and the interrelation between their welfare.

A more general expression for the MGNPV, for a society where people life-span is H periods (so there are H generations) and have a child when they are p years old, can be written as follows:

$$MGNPV = \sum_{g=-H}^{0} \left(\sum_{d=0}^{D} \beta^{d} NPV_{g+pd}\right) \tag{29}$$

Where d shows the propinquity of the descendant: for d = 0 it refers to individuals of generation g, for d = 1 to children, for d = 2 to grandchildren, and so on. D is the number of generations that are descendants of generation g and are affected by the project.

It should be noticed that the MGNPV does not apply a unique weight to any consumption in a certain moment of time as does the conventional discount factor, but it takes into account both the time preferences of the different generations and intergenerational altruism preferences to show appropriately the preferences of society about the different consumption flows.

Some Comments on Intergenerational Weighting and MGNPV

An appropriate consideration of the impacts occurring to future generations requires the application of an intergenerational weighting showing the preferences of society. This weighting depends both on the generational structure and the degree of altruism of the individuals, and clearly differs from the weight that involves the application of conventional discounting. Moreover, individual and social intergenerational weighting differ when the coexistence of more than one generation is taken into account.

A correct consideration of individual preferences about future consumption requires that instead of conventional NPV, the MGNPV should be maximized in project appraisal. This new evaluation method overcomes the arbitrariness of other methods, as it considers both the time allocation of own consumption, and the intergenerational allocation.

Individual and Social Optimality of Intergenerational Investments

In projects appraisal theory, it is usual to recommend the net present value (NPV) or the internal rate of return for determining the optimal level of investment, choosing between alternative projects, or simply, to indicate whether undertaking a specific project is profitable or not. These methods are based on the application of a constant time discounting to the impacts occurring in the future. Any impact that takes place in the very long run — like many environmental impacts — is given a negligible weight. The use of these methods therefore implies taking investment decisions, obviating what will happen to future generations, which usually leads to their interests being harmed. Furthermore, the application of a constant discounting does not follow any justifiable equity criterion and nor does it correspond to the true preferences of society. Future consumption is weighted according to the time preference of present generations. Conventional analysis thus implicitly assumes that present individuals have an infinite lifespan. However, it is obvious that part of the future flows of consumption will be enjoyed by other generations. As a consequence, it would be more appropriate to introduce the existence of different generations into the analysis and to consider the interrelations between them, thereby taking into account altruistic preferences, as we proposed in the previous section with the application of the MGNPV. The intertemporal allocation of individuals' consumption throughout their life-span should not be confused with the intergenerational allocation, which involves different generations. Both are part of the intertemporal allocation of resources and it would be inappropriate to study just one of its components, obviating the other.

The use of an overlapping generations model helps us to consider the efficiency of the prescriptions of conventional methods used for the evaluation of intergenerational investments, and to search which criterion should follow the public evaluation and management of these investments in order to achieve an adequate consideration of preferences. Nevertheless, as we have seen in section 4, the consideration of the interests of future generations might also require the application of certain equity criteria, such as the sustainability requirement.

An Overlapping Generations Model with Intergenerational Altruism

We use an overlapping generations model with intergenerational altruism with the following assumptions: the life-span of individuals is two periods, society is composed of two generations (old and young) of N individuals each; at the end of each period the old generation disappears and a new one appears, making the population consist of 2N individuals.

Following a non-paternalistic altruistic model, similar to the one in Barro (1974), we consider that individuals have the following utility function:

$$U_t = V_t + \delta U_{t+1} \tag{30}$$

where V_t is the life utility derived from the own consumption of an individual of generation t; $\delta = (1+R)^{-1}$ is the intergenerational discount factor, which we assume is between zero and one; U_{t+1} is the life utility function of his/her descendant.

The egoist component of the utility function is:

$$V_t = V(C1_t, C2_{t+1}) = u(C1_t) + (1+s)^{-1}u(C2_{t+1}) \tag{31}$$

where $(1+s)^{-1}$ is the discount factor that the individual applies to his/her second period consumption due to his/her time preference; $C1_t$ is the consumption of the individual in year t, when he/she is young; $C2_{t+1}$ is the consumption of the individual when he/she is old; $u(\cdot)$ is the function of satisfaction derived from consumption, which has the usual attributes of increasing utility of consumption $u'(\cdot) > 0$ and decreasing marginal utility of consumption $u''(\cdot) < 0$.

We can rewrite the utility function of an individual of generation t:

$$U_t = u(C1_t) + (1+s)^{-1}u(C2_{t+1}) + \delta U_{t+1} \quad (32)$$

In this type of model, there is a chain interrelating the utility functions of present generations with successive future generations. The welfare of any future descendant enters into the utility function of the individual who, in order to maximize his/her utility, has to maximize the next weighted sum of the satisfaction that his/her future descendants derive from consumption:

$$U_t = V_t + \delta U_{t+1} = V_t + \delta(V_{t+1} + \delta U_{t+2}) = V_t + \delta V_{t+1} + \delta^2 (V_{t+2} + \delta U_{t+3}) \quad (33)$$

So, we can express the utility function as:

$$U_t = \sum_{g=t}^{\infty} \delta^{g-t} V_g \quad (34)$$

In our analysis, we define social welfare as the sum of individual utilities.[13] As a consequence, the altruistic preferences of all the individuals of present society are taken into account in the social welfare function, because their utility functions include these preferences. If we assume that society at moment t is composed of two generations of N individuals, t-1 (old) and t (young), we have the following expression for the social welfare function:

$$W = \sum_{n=1}^{N} U^n_{t-1} + \sum_{n=1}^{N} U^n_t \quad (35)$$

Using (34) we obtain:

$$W = \sum_{n=1}^{N} \sum_{g=t-1}^{\infty} \delta^{g-(t-1)} V^n_g + \sum_{n=1}^{N} \sum_{g=t}^{\infty} \delta^{g-t} V^n_g \quad (36)$$

Where n is a dynastic label, so the n individual of generation t+1 is the heir to the n individual of generation t. In order to present a simpler notation, we omit this label in our following analysis.

[13] We adopt an extended version of utilitarianism where we include altruism in the utility function of individuals and not only egoist preferences as the pure utilitarian version does.

The Optimality of Intergenerational Investments

We study various cases of investments affecting more than one generation. We will compare the optimal decisions from individual and social perspectives, and analyze the differences between them and the prescriptions of conventional methods.

Factors Influencing on the Choice Between Own Consumption and an Investment that Yields Benefits to a Descendant

Starting with the altruistic utility function and the social welfare function considered above, we can make an initial approach to the factors influencing the determination of social and individual optimal decisions in intergenerational investments. We start by studying the case of the investment that an old individual gives to his/her descendant.

Firstly, we analyze the individual optimization problem. Looking at the total derivative of the utility function (32), we can deduce that an old individual in society at moment t will sacrifice his/her own consumption, in order to increase the consumption of his/her descendant, up to the following point:

$$(1+s)^{-1}u'(C2_t)\Delta C2_t = \delta[u'(C1_t)\Delta C1_t + (1+s)^{-1}u'(C2_{t+1})\Delta C2_{t+1}] \quad (37)$$

The left-hand side represents the cost in terms of utility involved in sacrificing consumption ($\Delta C2_t$) for investment. The right-hand side shows the increase in utility given by the increase in consumption of the descendant (flows $\Delta C1_t$ and $\Delta C2_{t+1}$) caused by the investment. We can see that in the investment decision, besides the magnitude of consumption flows, there are also other influential factors. These are the altruistic parameter δ, the marginal utilities of consumption, and the time discount. However, in the individual's optimal decision, the time preference applied to the flows enjoyed by the descendant is the time preference of the descendant and not the time preference of the investor, unlike usual decision rules such as the NPV, where the time discounting of present generations is used to discount all the flows of consumption generated by the investment.

We next analyze social optimization. Note that our social welfare function (36) consists of the sum of the utilities of all the individuals present at moment t in society - generations t-1 and t. In the investment of an individual of generation t-1 for his/her heir, all the effects modifying these utilities should be considered. The utility of the investor is affected by the utility of his/her descendant through the parameter δ, and the heir values his/her own consumption applying his/her own time preference.

The level of investment will be optimal when:

$$(1+s)^{-1}u'(C2_t)\Delta C2_t = (1+\delta)[u'(C1_t)\Delta C1_t + (1+s)^{-1}u'(C2_{t+1})\Delta C2_{t+1}] \quad (38)$$

As in condition (37), the marginal utilities of consumption and the altruistic parameter δ are quite relevant. However, the relative weights of the different consumption flows are clearly different from those of the individual condition because the consumption of the young individual has more weight in the social consideration. This consumption enters the social consideration not only due to the positive effect on his/her ascendant, through the altruistic parameter δ, but also due to its own valuation because his/her utility is part of the social welfare function. Furthermore, in both cases, the time preference is applied taking who enjoys the consumption into account.

In this analysis, we have seen some factors influencing social and individual optimal decisions in intergenerational investments that are not taken into account in conventional methods. These are the altruistic parameter δ, and distinguishing the beneficiary of the various flows of consumption when applying the time discount.

We next incorporate the individuals' budget constraints. We assume that the individuals have an inelastic labor supply in their first period of life, obtaining w_t in exchange, and transfer savings (s_t) to their second period of life at an interest rate r. We should stress that for the case of intangibles and non-market goods, it might not be possible to transfer consumption between periods at the market rate of interest. Moreover, even in the case of private goods, the imperfections of capital markets might prevent this from taking place in an efficient way. These problems suggest that in some cases we possibly cannot go farther than conditions (37) and (38). However, we will obviate these problems in this section, following the same steps as the usual theoretical analysis.

Investment of an Old Individual that Yields Benefits to His/Her Heir, both Present in Current Society

We first analyze an investment by which an individual of generation t-1 can transfer consumption flows to his/her descendant in the period of investment and in the one after it.

The budget constraints of the investor of generation t-1 are:

$$C1_{t-1} = w_{t-1} - s_{t-1} + Y1_{t-1}(b_{t-1}) \quad \text{(first period constraint)} \tag{39}$$

$$C2_t = (1+r)s_{t-1} - b_t + Y2_t(b_{t-1}) \quad \text{(second period constraint)} \tag{40}$$

where b_t are the resources that the investor of generation t-1 invests in the project benefiting his/her descendant; $Y1_{t-1}(b_{t-1})$ and $Y2_t(b_{t-1})$ are the consumption flows that receives from the investment b_{t-1} transferred by his/her parent. The investment b_t will yield the flows $Y1_t(b_t)$ and $Y2_{t+1}(b_t)$ to be enjoyed by his/her descendant.

The maximization program of the individual of generation t-1 is:

$$\max_{s_{t-1},\, b_t} \quad V_{t-1} + \delta V_t \qquad \text{s.t.} \quad (39) \text{ and } (40)$$

In the maximization program, we only need to consider the periods in which the decision has a direct impact — the investor's valuation of his/her own consumption (V_{t-1}) and of the consumption of his/her descendant (δV_t), as we assume that the descendant allocates his/her resources optimally.

The first order conditions are:

First order condition with respect to s_{t-1}:

$$u'(C1_{t-1}) = (1+s)^{-1}(1+r)u'(C2_t) \tag{41}$$

First order condition with respect to b_t:

$$(1+s)^{-1}u'(C2_t) = \delta[u'(C1_t)\, Y1_t + (1+s)^{-1}u'(C2_{t+1})Y2_{t+1}] \tag{42}$$

where $Y1_t = Y1_t'(b_{t-1})$ and $Y2_t = Y2_t'(b_{t-1})$ (we will henceforth use this notation)

Condition (41) shows the optimal level of saving. This is achieved when the marginal rate of substitution between consumption in period t-1 and consumption in period t is equal to the marginal rate of transformation.

Condition (42) tells us the optimal level of investment in the project that the descendant will enjoy. The left-hand side shows the cost in terms of utility of an additional sacrifice of consumption in the second period of the investor's life. The right hand side shows the increase in utility which yielded by the increase in consumption of his/her descendant (flows $Y1_t$ and $Y2_t$) caused by the marginal increase of the investment. The optimal level of investment depends not only on the consumption flows but also on the time preference, the parameter δ showing altruistic preferences, and the marginal utilities of consumption.

Given the assumptions about altruism ($\delta > 0$) and decreasing marginal utility of consumption ($u''(\cdot) < 0$), the investment in a descendant depends on his/her consumption level. This means that if there were the belief that the level of consumption increases with each successive generation, the individual would give a lower weight to the consumption of his/her descendants.

By combining (41) and (42) we obtain:

$$(1+r)^{-1}u'(C1_{t-1}) = \delta[u'(C1_t)Y1_t + (1+r)^{-1}u'(C^1_t)Y2_{t+1}] \tag{43}$$

$$(1+r)^{-1}u'(C1_{t-1}) = \delta u'(C1_t)\, [Y1_t + (1+r)^{-1}Y2_{t+1}] \tag{44}$$

which tells us that it is efficient to invest while the cost (left hand side) to the investor of a marginal increase in the investment is lower than the benefits (right hand side) that it yields. The consumption flows that the descendant receives are considered, and discounted them using the interest rate, but with the reference point of the beneficiary. They are then weighted using the altruistic parameter δ, which discounts the fact that consumption is enjoyed by the descendant and not the investor.

In order to facilitate comparison with the usual methods, and in the absence of better information, if we assume that $u'(C2_{t+1}) = u'(C2_t)$, which is equivalent to assuming that the different generations have a similar consumption level, the condition remains as follows:

$$(1+r)^{-1} = \delta[Y1_t + (1+r)^{-1}Y2_{t+1}] \tag{45}$$

This condition implies a weighting that notably differs from the one applied in usual evaluation methods, such as the NPV, in which the flows of consumption are weighted taking only the time from the moment of evaluation into account, without considering who enjoys the consumption. However, in expression (45), the consumption of the descendant is considered by applying the time discount from his/her point of view and weighting it by parameter δ, which shows how the investor values the increase in the utility of his/her heir.

In the case analyzed, taking t as the reference point for discounting, the NPV shows us that it is profitable to invest up to the next point[14]:

$$1 = Y1_t + (1+r)^{-1}Y2_{t+1} \qquad (46)$$

As we can see, the NPV does not consider who enjoys the consumption at all. It simply applies the time discounting from the point of view of present generations, as if these consumption flows were enjoyed by them.

Considering that our welfare function (37) is the sum of the utility functions of the individuals of society, the social maximization program for the considered investment is:

$$\max_{b_t} \quad V_{t-1} + (1+\delta)V_t \qquad \text{s.t. (39) and (40)}$$

The consumption of the descendant of generation t appears both for the consideration of his/her ascendant δV^t, and for his/her own consideration, because his/her utility is also part of the social welfare function.

First order condition with respect to b_t:

$$(1+s)^{-1}u'(C2_t) = (1+\delta)[u'(C1_t)\, Y1_t + (1+s)^{-1}u'(C2_{t+1})Y2_{t+1}] \qquad (47)$$

The condition regarding the optimal level of saving (41) remains unchanged. The social condition of the optimal level of investment (47) differs notably from the individual one (42). While in the individual case the weight given to the valuation of the consumption of the heir is δ, in the social case this weight is $1+\delta$. This occurs because both the positive effect that it causes on the investor (δ) and the consideration of the young individual as part of the social welfare function are taken into account.

By combining (41) and (47) we obtain:

$$(1+r)^{-1}u'(C1_{t-1}) = (1+\delta)[u'(C1_t)Y1_t + (1+r)^{-1}u'(C1_t)Y2_{t+1}] \qquad (48)$$

$$(1+r)^{-1}u'(C1_{t-1}) = (1+\delta)u'(C1_t)\, [Y1_t + (1+r)^{-1}Y2_{t+1}] \qquad (49)$$

which is different from (44). In this first case, we have observed that the inclusion of an overlapping generations model with intergenerational altruism can lead to divergence between individual and social optimal decisions. We have also observed that both decisions can differ from the decision indicated by the conventional application of discounting (condition (46)), which does not take into account who enjoys consumption but only considers the time distance from the moment of evaluation.

[14] For the sake of simplicity, in this analysis we assume that the discount rate used in the NPV is the interest rate r. The qualitative conclusions would not change with respect to the case that we used the time preferences.

Investment by a Young Individual that Yields Benefits to His/Her Descendant of the First Future Generation

In this case, we assume that the investor, belonging to generation t, can transfer consumption to his/her descendant through an investment project that yields benefits in the two following periods.

The budget constraints on the individual in generation t are:

$$C1_t = w_t - b_t - s_t + Y1_t(b_{t-1}) \quad \text{(first period constraint)} \quad (50)$$

$$C2_{t+1} = (1+r)s_t + Y2_{t+1}(b_{t-1}) \quad \text{(second period constraint)} \quad (51)$$

where b_t are the resources that the individual of generation t devotes to the project whose benefits are enjoyed by his/her heir; $Y1_t(b_{t-1})$ and $Y2_{t+1}(b_{t-1})$ are the consumption flows that he/she receives from the investment b_{t-1} transferred by his/her parent. The investment b_t yields the consumption flows $Y1_{t+1}(b_t)$ and $Y2_{t+2}(b_t)$ that his/her descendant receives.

The individual maximization program is:

$$\max_{s_t,\, b_t} \quad V_t + \delta V_{t+1} \qquad \text{s.t. (50) and (51)}$$

This includes the valuation that the individual of generation t has of the different flows of consumption that are directly modified by the investment and which affect his/her utility. These are the valuations of his/her own (V_t) and his/her descendant's (δV_{t+1}) consumption.

First order condition with respect to s_t:

$$u'(C1_t) = (1+s)^{-1}(1+r)u'(C2_{t+1}) \quad (41)$$

First order condition with respect to b_t:

$$u'(C1_t) = \delta[u'(C1_{t+1})Y1_{t+1} + (1+s)^{-1}u'(C2_{t+2})Y2_{t+2}] \quad (52)$$

The savings condition (41) remains unchanged. The left-hand side of the investment condition (52) shows the cost in terms of utility involved in an additional sacrifice of consumption during the first period of life. The right-hand side shows the increase in utility yielded by the increase in consumption of the heir caused by the increased investment in an additional unit.

Taking condition (41) one period forward, and substituting in (52) we obtain:

$$u'(C1_t) = u'(C1_{t+1})\delta[Y1_{t+1} + (1+r)^{-1}Y2_{t+2}] \quad (53)$$

The different consumption possibilities (considered using marginal utilities) as well as the altruistic parameter have a very important role in the individual's investment decision [15].

[15] The condition can also be expressed in function of the marginal utilities of consumption for the first period of life as we assume that the allocation of consumption along the life-span follows the optimality condition (41).

$$1 = [u'(C1_{t+1})/u'(C1_t))\delta(Y1_{t+1} + (1+r)^{-1}Y2_{t+2}] \quad (54)$$

If we assume that $u'(C1_{t+1}) = u'(C1_t)$:

$$1 = \delta[Y1_{t+1} + (1+r)^{-1}Y2_{t+2})] \quad (55)$$

This condition contrasts with the NPV criterion, where a constant time discount is applied without considering who enjoys the flows of consumption. The NPV states that the investment is profitable if the benefits are higher than the costs, which are both only weighted using the time discount factor. In this case, according to the NPV, it is profitable to invest until the next equality is achieved:

$$1 = (1+r)^{-1}Y1_{t+1} + (1+r)^{-2}Y2_{t+2} \quad (56)$$

In the investment under consideration, if $\delta > (1+r)^{-1}$ then the optimal decision of the individual in the presence of altruism is more generous to the heir than the NPV criterion. Taking into account that each period represents half of an individual's life span, it is not very daring to think that $(1+r)^{-1}$ is lower than δ.

The social maximization program is:

$$\max_{b_t} \quad (1+\delta)V_t + (\delta+\delta^2)\, V_{t+1} \qquad \text{s.t. (50) and (51)}$$

The investor's consumption appears twice because it affects the utility of his/her parent (δV^t), an older individual at the time the investment started. The consumption of the heir appears with two weights, one for the investor (δV_{t+1}) and other for the investor's parent ($\delta^2 V_{t+1}$) consideration. The valuation of the consumption of the ascendant of the generation t-1 (V_{t-1}) is not included in the program because it is not directly affected by this investment.

First order condition with respect to b_t:

$$(1+\delta)u'(C1_t) = (1+\delta)\delta[u'(C1_{t+1})Y1_{t+1} + (1+s)^{-1}u'(C2_{t+2})Y2_{t+2}] \quad (57)$$

$$u'(C1_t) = \delta[u'(C1_{t+1})Y1_{t+1} + (1+s)^{-1}u'(C2_{t+2})Y2_{t+2}] \quad (58)$$

The investment condition is identical to (52), so in this case, social and individual optimal levels of investment are the same and both differ from the NPV.

In this second case we have seen that even when a structure of overlapping generations and altruistic considerations is included in the analysis, it may be that individual and social optima of intergenerational projects are the same. However, it is not legitimate to generalize this result and to assert that social and individual optima are always the same, and even less so to assert that methods (such as the NPV), which only consider the flows of consumption by the time discount of present generations, lead to optimal decisions regarding the level of intergenerational investment, as some authors do with similar models[16]. In the investment in this example, the negative effect that the decrease in the consumption of the investor has on his/her ascendant (an individual present in today's society) compensates for the positive effect caused by the consumption of the heir of the investor (first future generation) in the social welfare function.

The result of this second case is due to the particular characteristics of the investment under consideration, but it cannot be generalized to all intergenerational investments. In fact, one of the main results of incorporating different generations and intergenerational altruism in the analysis is the divergence that might arise between social and individual optimal levels of intergenerational investments, as we have previously seen.

Investment of an Old Individual that Yields Benefits to His/Her Grandchild, Member of the First Future Generation

In the previous case, an investment taken by a young individual benefited his/her heir. Let us assume now that it is the old individual in t (generation t-1) who undertakes the investment enjoyed by his/her grandchild.

If the other circumstances are unchanged, the budget constraints of the investor in generation t-1 are as follows:

$$C1_{t-1} = w_{t-1} - s_{t-1} + Y1_{t-1}(b_{t-2}) \qquad \text{(first period constraint)} \qquad (59)$$

$$C2_t = (1+r)s_{t-1} - b_t + Y2_t(b_{t-2}) \qquad \text{(second period constraint)} \qquad (60)$$

where b_t is the investment that the individual of generation t-1 gives to his/her grandchild, member of generation t+1; $Y1_{t-1}(b_{t-2})$ and $Y2_t(b_{t-2})$ are the flows of consumption from the investment that the individual receives from his/her grandparent. The investment b_t yields the flows $Y1_{t+1}(b_{t-2})$ and $Y2_{t+2}(b_t)$ consumed by the grandchild of the investor.

The maximization program of the individual of generation t-1 is:

$$\underset{b_t,\ s_t}{\text{Max}}\ V_{t-1} + \delta V_t + \delta^2 V_{t+1} \qquad \text{s.t. (59) and (60)}$$

The project only modifies the consumption of the investor and his/her descendant of future generation t+1. We can thus obviate the valuation of the consumption of the descendant of generation t (δV_t) as well as the valuation of the consumption of the other descendants.

First order condition with respect to b_t:

$$(1+s)^{-1}u'(C2_t) = \delta^2 [u'(C1_{t+1})Y1_{t+1} + (1+s)^{-1}u'(C2_{t+2})Y2_{t+2}] \qquad (61)$$

The savings condition (41) does not change. Combining it with (61) we obtain:

[16] See e.g. Hultkrantz (1992), who identifies the social welfare function with the utility function of one member of the present generation, ignoring the point of view of the other present individuals. This

$$(1+r)^{-1}u'(C1_{t-1}) = u'(C1_{t+1})\delta^2[Y1_{t+1} + (1+r)^{-1}Y2_{t+2}] \quad (62)$$

$$1 = H\delta^2[Y1_{t+1} + (1+r)^{-1}Y2_{t+2}] \quad (63)$$

Where $H = [u'(C1_{t+1})(1+r))/u'(C1_{t-1})]$

Again, we obtain a condition depending on both the different marginal utilities of consumption and the altruistic parameter δ, which shows the positive effect of the descendant (grandchild) on the welfare of the investor. It is easy to see that this condition differs from the NPV. According to the NPV criterion, it is optimal to invest up to the point that the next equality is achieved:

$$1 = (1+r)^{-1}\, Y1_{t+1} + (1+r)^{-2}Y2_{t+2} \quad (64)$$

where the weights given to the different consumption flows do not consider who enjoy them but only when are they enjoyed.

The social maximization program is:

$$\underset{b_t}{\text{Max}}\ V_{t-1} + (\delta + \delta^2)\, V_{t+1} \qquad \text{s.t. (59) and (60)}$$

The weight given to the consumption of the descendant of generation t+1 is higher than in the individual program. This is because both the consideration by the parent (δV_{t+1}) and the grandparent ($\delta^2 V_{t+1}$) are included in the social valuation. Both of them are present in the society taking the decision and are thus part of the social welfare function.

First order condition with respect to b_t:

$$(1+s)^{-1}u'(C2_t) = (1+\delta)\delta[u'(C1_{t+1})Y1_{t+1} + (1+s)^{-1}u'(C2_{t+2})Y2_{t+2}] \quad (65)$$

If we introduce the savings condition (41) we get:

$$(1+r)^{-1}\, u'(C1_{t-1}) = u'(C1_{t+1})(1+\delta)\delta[Y1_{t+1} + (1+r)^{-1}Y2_{t+2}] \quad (66)$$

$$1 = H(1+\delta)\delta[Y1_{t+1} + (1+r)^{-1}Y2_{t+2}] \quad (67)$$

In the condition of the individual optimal level of investment (63) the weight given to the consumption of the individual of future generation t+1 with respect to the weight given to the sacrifice of an additional unit by the individual of generation t-1 was $H\delta^2$. In the social case (67) this weight is clearly higher, $H(1+\delta)\delta$. This is due to the fact that in the social valuation, the effects that the project causes on the utilities of all present individuals are included and not only the effects on the utility of the investor, as occurs in the individual case.

In general, the optimal social decision with respect to the level of intergenerational investment derived from a social welfare function including altruism is more generous to

leads, of course, to the coincidence between the social and the individual optimal investments considered in his work.

future generations than the optimal decision from the individual point of view. Furthermore, both of them differ from the NPV and very probably give more weight to consumption by future generations[17] because the positive influences of their consumption on the utility of their ascendants present in current society are taken into account. However, the NPV only considers the future through an arbitrary extension of the time preferences of present generations, thereby applying a negligible weight to future flows of consumption.

It is not necessary to repeat the entire analysis to affirm that the divergence is larger in the case of public goods, where the difference between the effects taken into account by an individual and the effects that enter into the social consideration are much larger.

Some Comments on Optimality in Intergenerational Investments

The use of an overlapping generations model has allowed us to introduce the existence of different generations and the interrelationships between their welfare, including altruistic preferences, into the analysis. We have thus been able to consider the problem of intergenerational allocation without avoiding the issue of the individuals' intertemporal allocation of consumption, as do most studies on the subject. Both issues are part of the intertemporal allocation of resources and must be taken into account in order to perform an appropriate analysis of it.

After considering different generations and intergenerational altruism, we have concluded that the optimal decisions of investment derived from individual and social preferences imply some factors that are not taken into account in usual calculations. Both in the individual and social decisions, the weight given to each consumption depends on who receives it and not only on the time discount of present generations, as is the case with conventional methods such as the NPV. Under the conditions of optimality derived in this section, the time discounting for the different flows of consumption is applied from the point of view of the individuals who receive them. These valuations are then weighted according to how they affect individual or social welfare, taking into account the altruistic preferences of individuals.

The optimal social decision regarding intergenerational investments is generally more generous with future generations than the individual one. This is due to the fact that in social decisions, all the effects that investments have on the utilities of the individuals within society are taken into account, and not only the effects on investors. Furthermore, both are very likely to be more favorable to future generations than conventional methods such as the NPV. These methods take into account all the flows of consumption, applying only the time discounting of present individuals and thus they ignore intergenerational allocation. It seems appropriate to modify the investment management and evaluation methods usually applied in intergenerational issues, as is the case with many environmental projects. In this regard, the application of the MGNPV proposed in the previous section would lead to the weighting of future consumption, taking into account the preferences regarding the consumption of future generations. This will help to avoid the negligible weight usually given to future generations. The application of conventional discounting is misleading with regard to the true preferences

---

[17] In this specific case, assuming that the different generations have the same marginal utility of consumption, this is true if $(1+\delta)\delta$ is higher than $(1+r)^{-1}$.

of society and could lead to investment well below the optimal level in many environmental and infrastructure projects.

CONCLUSION

The set of economic instruments for internalizing externalities and providing collective goods are effective to a more or less satisfactory extent for solving the environmental problems affecting present generations. However, these instruments are inappropriate for considering the externalities that present generations cause to future generations and for preserving intergenerational goods.

Obviously, if the world ended with present generations, problems such as climate change or the loss of biodiversity would merit much less attention. The reason for this is simple - the costs to present generations are relatively small, while it is to be expected that they are very important for future generations.

It should be noted that a necessary condition for internalizing intergenerational externalities is to have a system which allows future generations to have a part of the property rights and to be able to manage and negotiate with them without any limitations other than those governing the behavior of the present generation. For the moment, there is no institution which could transfer property rights to future generations. Efforts towards environmental management according to conventional economic theory therefore only work to the benefit of the present generation, in as far as they do not consider the interests of future generations.

This is an economic problem, but only in part, because any solution to it requires a new institutional framework. It is necessary to create new institutions which allow future generations to

a. Have their rights over natural resources and the environment recognized.
b. Be able to negotiate their property rights in the market with the same rights and possibilities as present generations.
c. Have access to administrative and law systems
d. Have a voice and vote in the political system.

To conclude, we feel that the main problem is the conflict of interests between present and future generations. The consequences of this conflict from a distributive perspective have been studied (see e.g. Lind, 1995) However, the impact from an allocative point of view have been studied to a lesser extent [18] and for this reason, we have devoted the second part of our study to this important problem.

Research has focused on the search for a satisfactory solution to the consideration of future generations in project and policy appraisal. Conventional cost-benefit analysis loses its legitimacy in the intergenerational context. An appropriate treatment of intergenerational problems needs to overcome the strong limitations of conventional economic analysis, such as the very unequal distribution of rights or the infinite substitution possibilities it assumes. The paper has presented an alternative evaluation process consistent with the sustainability

[18] In Pasqual and Souto (2003) the relationships between present and future generations are analyzed, taking into account both the distributive and the allocative effects.

requirement in which the right of future generations to enjoy a non-diminished resource capacity is recognized. The consideration of preferences over the consumption of future generations has led us to recommend a new valuation method, the multigenerational net present value, for considering the impact on each generation by means of the correct intergenerational weight. Finally, analysis of various possibilities for intergenerational investments has shown that conventional methods do not take into account preferences for the consumption of future generations in an appropriate manner, which reinforces the need to apply the new methods in order to consider the consumption flows in each generation adequately.

REFERENCES

Barrett, C.B. (1996). Fairness, stewardship and sustainable development. *Ecological Economics,* 19, 11-17.

Barro, R.J. (1974). Are Government Bonds Net Wealth? *Journal of Political Economy*, 82, 1095-117.

Bator F. M. (1958). The anatomy of market failure. *Quarterly Journal of Economics*, 72, 358.

Baumol, W. (1952). Welfare State Economics and the Theory of State. *Cambridge: Harvard University Press.*

Bellinger, W. K. (1991). Multigenerational value: modifying the modified discount method. *Project Appraisal*, 6, 101-108.

Bird, P., J. W. N. (1987). The transferability and depletability of externalities. *Journal of Environmental Economics and Management,* 14, 54-57.

Bromley, D.W. (1989). Entitlements, missing markets, and environmental uncertainty. *Journal of Environmental Economics and Management*, 17, 181-194.

Broome, J. (1992). Counting the Costs of Global Warming. *Cambridge: White Horse Press.*

Buchanan, J., and Craig, W. (1962). Externality. *Economica*, 29, 374.

Coase, R. H. (1960). The problem of social cost. *Journal of Law and Economics*, 3, 1-44.

Collard, D. (1981). Altruism and Economy. *Bath: The Pitman Press.*

Costanza, R. (1994). Three general policies to achieve sustainability. In A. Jansson, M. Hammer, C. Folke, and R. Costanza (Eds.), *Investing in Natural Capital: the Ecological Economics Approach to Sustainability* (pp. 392-407). Washington DC: Island Press.

Costanza, R., and Perrings, C. (1990). A flexible assurance bonding system for improved environmental management. *Ecological Economics*, 25, 55-57.

Doeleman, J. A. (1988). Social Determinism, technology and economic externalities. In C. Tisdell, and P. Maitra (Eds.), *Technological Change, Development and the Environment: Socio-Economic Perspectives* (pp. 298-321). London and New York: Routledge.

Howarth, R. B., and R. B. Norgaard. (1995). Intergenerational Choices Under Global Environmental Change. In D.W. Bromley (Eds.), *Handbook of Environmental Economics* (pp. 111-138). Oxford: Blackwell Publishers.

Howarth, R. B. (1997). Sustainability as opportunity. *Land Economics*, 73, 569-579.

Hultkrantz, L. (1992). Forestry and the bequest motive. *Journal of Environmental Economics and Management,* 22, 164-77.

Katz, M. L., and Shapiro, C. (1985). Network externalities, competition and compatibility", *American Economic Review*, 75, 424-440.

Markandya, A., and Pearce, D. (1988). Sustainable future. Natural environment and the social rate of discount. *Project Appraisal*, 3, 2-12.

Mishan, E. J. (1971). The postwar literature on externalities: An interpretative essay. *Journal of Economic Literature*, 9, 2-3.

Neumayer, E. (1999). Weak versus Strong Sustainability. Exploring the Limits of Two Opposing Paradigms. *Cheltenham: Edward Elgar*.

Nijkamp, P., and Rouwendal, J. (1988). Intergenerational discount rates in long-term plan evaluation. *Public Finance/Finances Publiques*, 43, 195-211.

Norton, B. G. (1995). Evaluating ecosystem states: two competing paradigms. *Ecological Economics,* 14, 113-27.

Ocaña, C., and Pasqual, J. (1994). Environmental costs of residuals: a characterization of efficient tax policies. Revista Española de Economía, *Special Issue: Recursos naturales y medio ambiente*, 51-59

Padilla, E. (2002). Intergenerational equity and sustainability. *Ecological Economics*, 41, 69-83.

Page, T. (1983). Intergenerational justice as opportunity. In D. MacLean, and P. G. Brown (Eds.), *Energy and the Future*. Totowa, NJ: Rowman and Littlefield.

Page, T. (1997). On the problem of achieving efficiency and equity, intergenerally. *Land Economics,* 73, 580-596.

Pasqual, J. (1999). La Evaluación de Políticas y Proyectos. Barcelona: Icaria Editorial and Universitat Autònoma de Barcelona.

Pasqual, J., and Riera, P. (2004). Underground land values. *Land Use Policy*, forthcoming.

Pasqual, J., and Souto, G. (2003). Sustainability in natural resource management. *Ecological Economics*, 46, 47-59.

Pigou, A. C. (1920). *The Economics of Welfare*. London: Macmillan.

Price, C. (1993). *Time, Discounting and Value*. Oxford: Blackwell.

Price, C. (1996). Discounting and project appraisal: from the bizarre to the ridiculous. In C. Kirkpatrick and J. Weiss (Eds.) *Cost–Benefit Analysis and Project Appraisal in Developing Countries*. Cheltenham: Edward Elgar.

Samuelson, P. A. (1954). The pure theory of public expenditure. *Review of Economics and Statistics,* 36, 387-389.

Samuelson, P. A. (1958). An exact consumption-loan model of interest with or without the social contrivance of money. *Journal of Political Economy*, 66, 467-482.

Thompson, E. (1966). A Pareto optimal group decision process. In G. Tullock (Ed.) Papers on Non-Market Decision Making (pp. 133-140). Charlottesville: University of Virginia.

Uhlaner, C. J. (1989). Democracy and participation: incorporating sociability into a rational action. *Public Choice*, 62, 253-285.

World Commission on Environment and Development (WCED). (1987). *Our Common Future*. Oxford: Oxford University Press.

INDEX

A

B

C

D

E